TRAINING AND PER
in Difficult Enviro

A GUIDE FOR COMPETITIVE ATHLETES

Christmas 2006

To Stewart & Anna.
With love from
Evan

speedo

TRAINING AND PERFORMANCE
in Difficult Environments

A GUIDE FOR COMPETITIVE ATHLETES

Stanley Grant and Evan Lloyd

THE CROWOOD PRESS

First published in 2006 by
The Crowood Press Ltd
Ramsbury, Marlborough
Wiltshire SN8 2HR

www.crowood.com

British Library Cataloguing-in-Publication Data
A catalogue record for this book is available from the British Library.

ISBN-10 1 86126 881 5
ISBN-13 978 1 86126 881 5

Acknowledgements
The authors wish to thank the following people for their help in the preparation of this book: Tom Aitchison; Dr John Bradley; Charles Campbell; Ian Donnelly; Trevor Graham; LeeAnn Hampson; Dr Niall MacFarlane; Dr John MacLean; Andrew Murray; Paul Paterson; Alan Ramage; Dr Brian Walker; Dr Martin Watt.

Disclaimer
Please note that the authors and the publisher of this book do not accept any responsibility whatsoever for any error or omission, nor any loss, injury, damage, adverse outcome or liability suffered as a result from practising, or applying, the principles, ideas, techniques and/or following the instructions/information described in this publication. Competing and training in any environment can be dangerous and may be too strenuous for some readers to engage in safely and it is essential that a doctor be consulted prior to undertaking any training.

Frontispiece: Reproduced with permission from Andrew Murray, The National Stadium, Sports Health & Injury Clinic, Hampden Park, Glasgow

Typeset by Florence Production Ltd, Stoodleigh, Devon
Printed and bound in Spain by GraphyCems

Contents

Introduction

Every athlete or sports person wants to perform at his or her best in competition and this is best achieved by meticulous preparation. Individuals may embark on training for endurance and speed, make a detailed study of diet and nutritional needs, and give attention to psychological issues, while teams will look in depth at strategy and team moves. While all these factors will have a major bearing on an athlete's performance, one of the most important things to remember is that sport is always carried out within a specific environment. This environment may affect the participant in many ways, all of which could mean that, despite all the training and preparation, the athlete will not reach his or her potential peak of performance.

Most people are used to performing in their home environment but, even in a familiar situation, they may not be aware of all the potential problems and hazards. The 'home' environment will vary depending not only on whether the activity is taking place outdoors or indoors, but also on the environment itself, which will change, certainly during the year, but also sometimes within the same day. There are many aspects of the environment that may adversely affect performance and increase the risk of injury or ill health. The athlete who has formulated strategies to counter or limit problems associated with hostile environments may beat a 'better' athlete who has failed to prepare for the adverse conditions. Some of the items discussed may seem to be of minor relevance, but it is vital to remember that the difference between winning and coming second is often minuscule – a fraction of a centimetre or a few hundredths of a second.

The environment can affect sportsmen and women not only through obvious factors such as temperature, time zone changes and altitude, but also through other variables that may be present in a particular geographic location. These include pollution of air and water, and insects. Not only may an insect bite impair performance directly (for example, when it occurs on an ankle where it is rubbed by the shoe), but a bite, or a failure of hygiene, may also introduce one of a range of diseases. Measures can be taken to avoid many of these harmful effects. Long before they set off, athletes who travel for competition, matches or training should consider in great detail the potential hazards of the destination venue. They should find out what precautions can be taken before leaving (for example, immunizations), what safety measures should be adopted at the new location, what treatments are possible, and what should be checked on their return (remembering that some health problems acquired abroad may not show clinical features immediately).

This book is for sportsmen and women, coaches and anyone else involved with athletes who will perform and train in a variety of environments. It identifies a number of environmental stressors, and outlines how an athlete's performance and health may be affected. It explains the difficulties associated with exercise in heat or cold, or at altitude, long-distance travel and air pollution, and gives practical guidelines for strategies to counter potential problems. On some subjects, it is impossible to give advice that applies across the board; for example, with malaria, the list of countries at risk, and the information on the appropriate preventive drugs change quite regularly. Addresses are therefore given to point the reader towards the most up-to-date advice. Guidance for those with some pre-existing medical conditions is also given, as well as information on infections and strategies to prevent and counter them.

The background to the problems associated with changes in environmental temperature is covered in a chapter on thermoregulation, or how the body gains and loses heat, while Appendix I, on some aspects of exercise physiology, will assist with the non-scientific reader's understanding of a potentially difficult topic. Appendix II provides details of some human responses to altitude, using physiological terms and offering a deeper study of altitude physiology; and some aspects of fluid balance in the body are covered in Appendix III. While it is not essential to read the appendices, they will provide a useful reference before, during or after reading the individual chapters.

This book was written to be understood by the lay reader, but readers will also benefit from having access to some of the sources used by the authors. While not all sources have been referenced, the text is based on the scientific literature. Clearly, some readers may not agree with the authors' interpretation of the research, but guidelines are given where possible, in order to aid the performer and coach to shape strategies that will limit possible performance decrements in any given environment. With a view to helping lay people as well as those with more scientific knowledge, whenever a technical term is used, an explanation is given in the text and/or in the glossary.

Further information may be sought in the list of websites at the end of the book, as well as in the Further Reading section. Many of the articles from scientific journals are now available online.

CHAPTER 1

Thermoregulation

THE CORE AND THE SHELL

Temperature within the body is by no means uniform. For convenience, the body can be divided into the 'core' and the 'shell'. The core comprises the brain, chest, abdominal areas and some skeletal muscles. The shell refers to the external parts of the body, and its thickness varies. Deep (core) body temperature is normally maintained at around 37°C. This constancy is achieved as a result of a balance between the rate of heat production (predominately from chemical reactions that take place in the cells, and muscle contractions known as metabolic heat generation) and the rate of heat loss. While the core temperature is tightly regulated, within a narrow range, the temperature of the shell – particularly of the shell's outer layer, the skin – can vary widely.

The size of the shell (*see* Figure 1) varies depending on the environmental temperature. In a very cold environment, it is important to defend the body against heat loss and the shell may extend some centimetres below the skin. However, in very hot conditions it may be as little as 1cm thick. Thus, alteration of the temperature of the superficial tissues (the skin) plays a role in the regulation of the core temperature. Clearly, an external part of the body (the shell) can have a much lower temperature than the inside of the body. The shell responds to changes in the external environment while the core temperature remains relatively constant. Figure 1 highlights the fact that the core is much smaller in the hypothermic situation compared with that of hyperthermia or heat stroke. In a cold environment the leg skin temperature (shell) may be 20°C lower than the core temperature. For example, in a cold environment (–10°C) we have recorded a core temperature of 37°C while a calf skin temperature was 15°C. Therefore, the term 'body temperature' is inappropriate. It is more informative to refer to core temperature or skin temperature.

The core and the shell

- The core comprises the brain, chest, abdominal areas and some skeletal muscles.
- The shell refers to the external parts of the body.
- Deep (core) body temperature is normally maintained at around 37°C.
- The temperature of the shell – particularly of the shell outer layer, the skin – can vary widely. The size of the shell and shell's thickness varies depending on the environmental temperature.

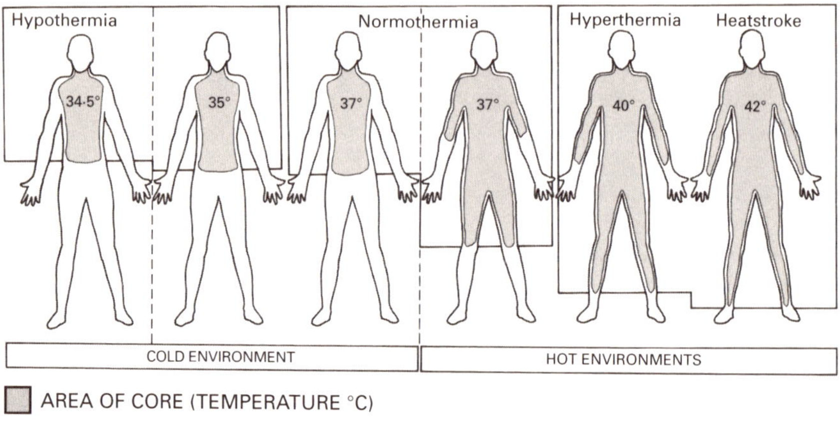

Fig 1. The core and the shell.

A rough indication of the relationship between core temperature and shell size. Cold and hot environments may be responsible for illnesses and injuries and may affect performance. (Originally published in Lloyd, E. L. (1994), ABC of Sports Medicine: Temperature and Performance I: Cold, *British Medical Journal*, 309, 531–4. Reproduced with permission from the BMJ Publishing Group)

Core and skin temperature

Temperature within the body is not uniform. At times, the skin temperature in some sites may be 20°C less than the core temperature.

While the core (deep body) temperature in humans is normally around 37°C, there is diurnal variation, with core temperature being lowest in the morning and highest in the evening. The location of the measurement site will influence the core temperature value. The oesophageal temperature (in the oesophagus, the tube that connects the mouth with the stomach) is normally higher than rectal, tympanic (ear) and oral (mouth) temperatures. The oesophageal temperature responds faster to changes in core temperature than the rectal temperature. Oral temperatures can be influenced by the temperature of recently drunk fluids or ingested food, or even by heavy breathing of very cold air with the mouth open. Oral and tympanic temperatures may be influenced by environmental temperatures, for example, in the case of immersion in cold water.

Comparatively small deviations from a core temperature of 37°C can be life-threatening. For example, high-intensity exercise for a prolonged period in heat may lead to hyperthermia (overheating of the body). A core temperature greater than 41°C may lead to convulsions; a core temperature of 43°C will probably result in death. If the core temperature is less than 35°C, the individual has hypothermia. The degree of physiological malfunction associated with hypothermia is related to changes in core temperature but it will vary from person to person. Disturbances to many systems of the body may occur before the core temperature falls to hypothermic levels. For example, effects on the central nervous system can promote mistakes, slurred speech and irrational

Hyperthermia and hypothermia

- Hyperthermia is overheating of the body.
- Hypothermia, cooling of the body, is diagnosed when the core temperature is less than 35°C.

behaviour, such as taking clothes off in the cold. As core temperature decreases to around 33°C, shivering slows down and finally stops. Cardiac arrhythmias (irregular heartbeats) often occur at around 28°C.

Maintaining the core temperature at around 37°C is achieved in humans via the thermoregulatory centre (thermostat) of the body, in the hypothalamus (the part of the brain that influences the control of some of the internal workings of the body). Thermal receptors provide the hypothalamus with temperature information. This information is processed and, depending on how the information compares with a set point (37°C), the hypothalamus may bring into play thermal effectors (*see* below) to adjust the core temperature to the set point value. The thermal receptors in skin (peripheral thermoreceptors) are sensitive to changes in skin temperature and act as an early warning system that changes are needed, so that the core temperature can be maintained within desirable limits. The central thermoreceptors in the hypothalamus detect changes in the temperature of the arterial blood.

Individuals are not conscious of the events surrounding thermoregulatory control in the hypothalamus, but they can certainly perceive warm and cold sensations. In cold situations, a number of strategies may be adopted in order to keep warm: putting on clothes, exercising or possibly curling up. In heat, a person may seek shade or spread himself out in an attempt to stay cool. However, the body is much more aware of changes in environmental temperatures rather than absolute values. For example, one of the authors felt warm and took off the outer layer of clothing on going into a room at 0°C whereas he started shivering on going into an area kept at 21°C. In the first situation he had just come out of a room kept at –30°C, with additional wind, whereas in the second he had left an outdoor temperature of 35°C.

HEAT BALANCE

The body responds to heat and cold in a number of ways in an attempt to prevent major disturbances to normal core temperature.

Heat Production

The amount of heat generated when a person is lying totally at rest is called the Basal Metabolic Rate (BMR); this is the energy expended by the body in a given time during rest in fasted conditions. Additional heat is generated by working muscles and, in the cold, further heat can be generated via the skeletal muscles as a result of shivering when two sets of muscles contract against each other. For example, the muscles in the front and the back of the thigh contract and produce heat, which can elevate the core temperature. Heat production

Fig 2. Differences in the volume of air breathed out over a 5-minute period between shivering and non-shivering.

(Reproduced with permission from Dr Martin Watt, Accident and Emergency, Monklands Hospital, Airdrie)

through shivering can be increased four- or five-fold above resting values. Increased heat production requires increased oxygen uptake (VO_2), which may rise by up to 400 per cent during shivering.

In the test illustrated in Figure 2, the subject had been lying in a chamber for an hour at a temperature of –10°C and a wind speed of 10mph. VO_2 and the volume of air breathed in and out per minute were measured over 5-minute durations at various times throughout the test. The volume of air in the bag in the subject's left hand was much larger than that on the right. The latter was measured at the start of the trial when the subject was not shivering and the former near the end of the test when the subject was shivering violently. The subject was visibly shaking on a regular basis, with some periods when no movement was apparent. The VO_2 and the volume of air breathed out per minute were much higher in the left-hand bag, during the period when there was shivering.

Heat Loss

The temperature gradient between the skin and the environment will influence the heat transfer. When the temperature of the skin is markedly higher than that of the environment, there will be a pronounced heat loss; when the skin and environmental temperatures become similar, heat loss is much reduced. When the environmental temperature is greater than skin temperature, the body gains heat. Normally, an individual's body is warmer than the surrounding environment and there is net loss of heat to the cooler environment.

Heat is generated by muscle activity and, since muscles have direct contact with tissue and skin, that heat can be lost via conduction as it is transferred to the skin surface by direct contact, and thence to the environment. Heat is also transferred by the circulation. The smooth muscles of the arterioles (the very small blood vessels between the larger arteries and the capillaries, which actually distribute oxygen and nutrients to the tissues) influence heat loss by adjusting blood flow. In a cold environment, blood flow to the skin is reduced because the smooth muscles in the arterioles contract and shut down the blood flow to the skin. As a consequence, less blood and heat are carried from the core to the periphery, and the temperature gradient between the skin and environment is reduced, resulting in a less pronounced heat loss. When the blood flow to the skin has been 'shut down', the thickness of the layer of fat under the skin (subcutaneous fat) has a bearing on heat loss. The greater the thickness of this layer, the less heat is lost. Conversely, in a hot environment there is an increase in blood flow to the skin. In this way, the body attempts to carry heat from the core to the skin where heat is lost to the environment, as long as the environmental temperature is less than the skin temperature.

Sweat secretion from around 3 million sweat glands results in cooling of the skin through evaporation. When 1ltr of water vaporizes from the skin, 580kcals of heat are removed from the body. However, in humid conditions, evaporative heat loss is inhibited. Relative humidity is described as the ratio of water in ambient air in relation to the maximum amount of moisture that the air is able to carry for a given ambient temperature. For example, 70 per cent relative humidity indicates that the ambient air has 70 per cent of the air's moisture-carrying capacity that it is capable of carrying at that temperature. With sweating, the skin becomes wet, resulting in an increase in vapour pressure (humidity) immediately adjacent to the skin. If the difference in vapour pressure between the skin and the environment decreases (as it does in high-intensity exercise in a humid environment), evaporative heat loss is limited and the exerciser is more likely to overheat than in less humid conditions. Higher environmental humidity results in less evaporation and therefore less heat loss through evaporation. Thus, evaporation of sweat is much less in a tropical climate than in a desert environment. It must be stressed that it is evaporation of sweat that results in cooling and not simply sweat production. Sweat that rolls off the skin does not promote heat loss.

Evaporative heat loss

- If the difference in vapour pressure between the skin and the environment becomes small (as it does in high-intensity exercise in a humid environment), evaporative heat loss is limited and the exerciser is more likely to overheat than in less humid conditions.
- It is evaporation of sweat and not simply sweat production that results in cooling.

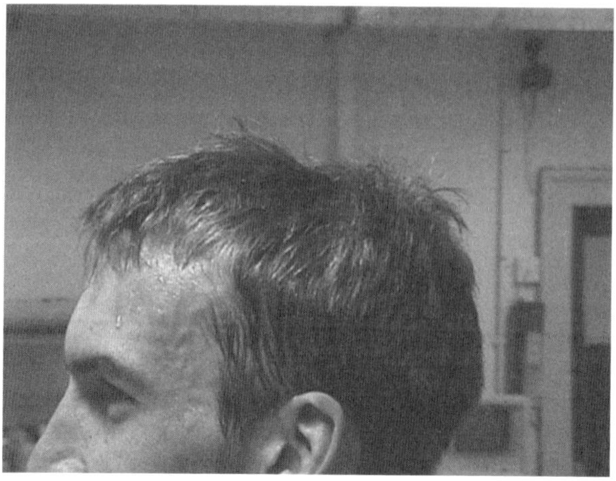

Fig 3. Sweat rolling off skin.

The sweat on the athlete's skin has to evaporate before there will be any cooling effect. Sweat on the skin or rolling off the skin does not contribute to heat loss. (Reproduced with permission from Mr Paul Paterson, University of Glasgow)

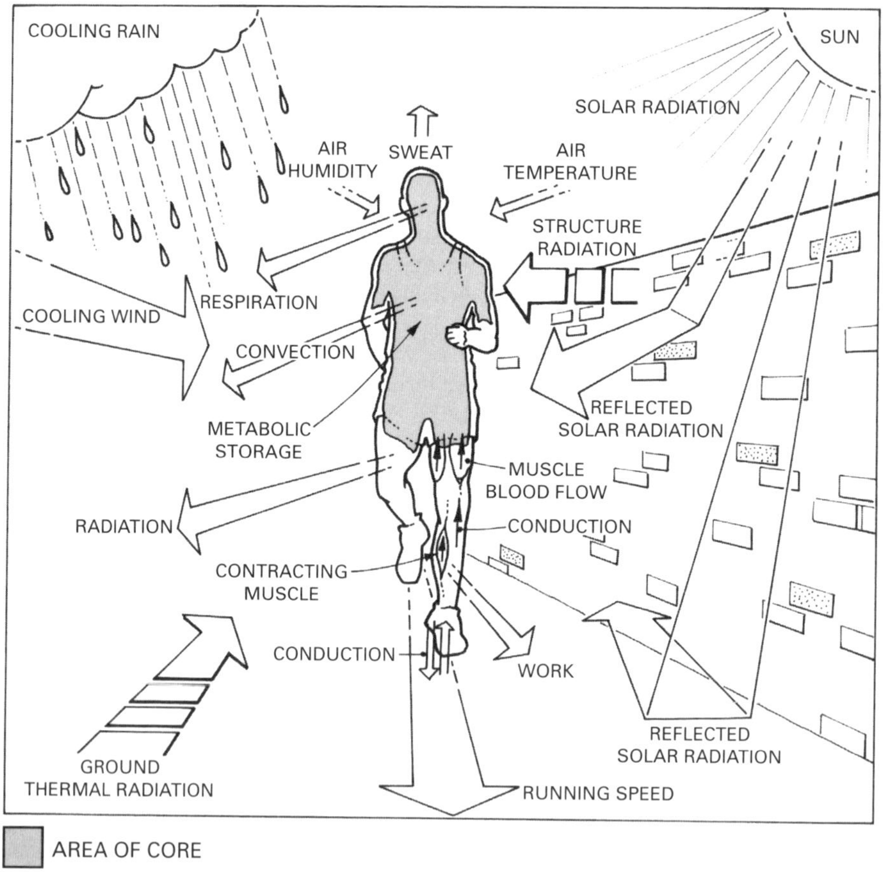

Fig 4. Schematic diagram showing some of the possible heat-exchange mechanisms during exercise.

Heat exchange mechanisms

Next time you have a bath you can ponder heat exchange mechanisms. The body will gain radiant heat from the radiator if it is on, and possibly from any sun shining through the window. A foot immersed in warm bath water will gain conductive heat as it is warmed by the water. Conversely, placing a foot in an icy river will result in conductive and convective heat loss. If you leave the bathroom door open, the resulting movement of cold air from outside will result in convective heat loss. If you move to a bedroom, you will experience radiative heat loss if the surfaces (walls, furniture) are cooler than your body.

Blood flow to and from the skeletal muscles is greatly increased during exercise. Heat produced by the working muscles is carried to the core by the blood by convection via the veins and then to the skin by conduction and convection, via the arterial blood flow. This results in an increase in core and skin temperatures.

When an individual exercises at high intensity in heat, the cooling from radiation, convection and conduction is often inadequate to prevent an increase in core temperature. Thus, heat loss from evaporation is the main means by which the exerciser may limit heat gain. In prolonged high-intensity exercise in certain environmental conditions, the amount of heat produced (the greater the exercise intensity, the greater the heat production) may be so great that the heat exchange mechanisms may not be able to cope and the body will gain heat. This situation occurs frequently when it is hot and there is high humidity. The high humidity limits evaporative heat loss. In calm conditions, or if an athlete is moving in the same direction and at the same speed as the wind, there will be little convective heat loss. If the athlete is training or racing under a cloudless sky at around noon in a low-latitude location, there will be considerable radiative heat gain. Given these conditions, the endurance exerciser is likely to experience heat problems.

Summary of heat exchange mechanisms and exercise examples

- **Conduction:**
  - direct contact between body part and other surface.

  Examples,
  - sitting on substitutes' bench;
  - holding a ski pole.

- **Convection:**
  - movement of water or air;
  - movement of blood to other body areas with a different temperature from the blood.

  Examples,
  - running into a head wind, which has a greater cooling effect than a tail wind, or standing in calm conditions with no wind;
  - movement of air in layers of clothing as a result of rhythmical limb movement;
  - movement of blood warmed by working skeletal muscles to body area at lower temperature than the blood.

- **Radiation:**
  - direct transfer of heat across a space by electromagnetic waves (no need for the surfaces to touch as in heat gain from a fire or the sun).

  Examples,
  - radiant heat gain from surfaces such as a synthetic pitch or a tennis court;
  - radiant heat gain from exposure to direct sunlight, particularly at low latitudes around noon – perhaps running in Singapore.

- **Evaporation:**
  - liquid changes to a gas and results in heat loss;

  Examples,
  - evaporation rate of water, wet clothes or sweat, is increased by wind;
  - evaporation of sweat cools the athlete;
  - evaporation is high in 'dry' desert heat (for example, in the Marathon des Sables, a race in the Sahara Desert) but limited in high humidity as in a tropical heat (for example, at the 1994 Football World Cup in Florida), where there is a decreased water vapour gradient between the skin surface and the air.

CHAPTER 2

Exercise in Heat

ATHENS OLYMPICS 2004

In August 2004, British athlete Paula Radcliffe was the fastest woman marathoner in the world. Expectation in Britain was high as she prepared for the Athens Olympic Marathon, which was run over the original marathon course. Paula had an impressive recent history in marathons. She was first in the 2002 London Marathon in 2 hours 18 minutes 56 seconds (2:18:56 – a UK and European best); she won the 2002 Chicago Marathon in 2:17:18 (a world best); and she won again in 2003 in London, in 2:15:25 (a world best). Such remarkable consistency, allied to her outstanding performances, made her a clear favourite for Olympic gold.

The Olympic marathon was run in very hot conditions, with an ambient temperature of about 33°C. Paula Radcliffe stayed with the leading group for the first 15 miles but was unable to respond to the Japanese, Noguchi, when she pulled away at 16 miles. Noguchi won in a time of 2 hours 26 minutes and 20 seconds (over 10 minutes slower than Paula's best time). Paula persevered but finally stopped and sat at the side of the road, a forlorn figure.

Why did Paula encounter such serious problems? Why did an athlete who had run two minutes faster than her Athens competitors fail to win; indeed, why did she not even finish? It could be that Paula simply had a bad day at the office or she may have had some underlying illness. Some time after the race, she did state that she had been below par before the event. However, the scientific literature indicates that endurance performance is compromised in heat. It is therefore to be expected that athletes will perform less well in heat in endurance events.

Did the conditions have a proportionately greater effect on Paula than on her competitors? In a post-race interview, she did not blame the conditions, which were the same for everyone. The BBC commentary team of Steve Cram and Colin Jackson highlighted the fact that the winner was certainly more accustomed to heat than Paula. David Bedford, the one-time 10k world record holder, is of the opinion that it is not possible for athletes from temperate climates to win a marathon in the sort of conditions that prevailed in Athens. Reports would suggest that Paula certainly considered the problems that heat might present. She did perform a warm-up with an ice-cooled jacket – to help reduce the increase in core temperature that would have been the result of the warm-up – but apparently she was unable to wear it for the prescribed hour before the race. Dehydration has been shown to reduce endurance performance. Paula is reported to have drunk an electrolyte drink at every feed station. Is it possible that she had incipient hyponatraemia?

Paula normally races from the front. Given the hostile conditions, and in recognition of the fact that a diminished performance time was inevitable, she may have reduced her race pace, and changed her tactics. Exercise in heat is associated with a greater perception of effort compared with cool conditions. Thus, Paula would have felt less comfortable at her 'normal' race pace. Paula did stress that her training regime had been structured to take the Greek weather conditions into consideration. Paula had heat acclimatized. Heat acclimatization enhances performance in heat but is unlikely to allow athletes to perform as well as they would in temperate conditions.

Paula Radcliffe's experience throws up a number of important questions:

- Was a warm-up advisable or necessary before the Athens marathon?
- Should athletes try pre-cooling strategies before endurance events in the heat?
- Is heat acclimatization of benefit for the endurance athlete who will compete in heat?
- Is a reduction in pace advisable in hot conditions?
- What should the endurance athlete drink and why?

While it may seem obvious that endurance athletes are likely to have problems in heat, it should be stressed that other categories of athlete may experience adverse effects when living and exercising in heat and high humidity. Athletes who travel to a hot and very humid location may experience difficulties simply by living and training in the heat. Care must be taken to ensure that the athlete, whatever the sport, does not become dehydrated as a result of living in the unaccustomed heat of a new location.

Players of games such as soccer, rugby and hockey sometimes take part in prolonged intermittent exercise in hot conditions, and are susceptible to heat problems. Soccer players have been shown to have an average sweat loss of 2193ml (range – 1672–3138ml) after 90 minutes of game related training in very hot conditions. Sprint athletes, tennis players and short-duration endurance athletes may be involved in heats and finals in the same day, and many athletes train more than once a day. As a result, they may be dehydrated due to a previous exercise bout, and consequently under-perform.

Exercisers should be able to evaluate 'hot conditions'. Scrutiny of the competitive climatic conditions is advised long before competition so that weather conditions can be evaluated and strategies devised to help minimize the problems. The athlete should be aware of possible problems while exercising in heat and high humidity, and give consideration to tactical adjustments, pay particular attention to fluid intakes, and devise a heat-acclimatization strategy. The athlete who has strategies to cope with the environmental challenges may beat a 'better' athlete who fails to prepare for the hostile environment. Athletes who do not attend to strategies to lessen the adverse consequences of exercising in heat are guilty of neglect.

HOW HOT IS HOT?

The higher the ambient temperature the smaller the gradient for heat loss between the skin and the surrounding air, resulting in reduced heat loss. Heat gain takes place when the ambient temperature is higher than the skin temperature. The higher the humidity, the less efficient evaporation is and therefore

> **Exercise in heat**
>
> - Heat gain takes place when the ambient temperature is higher than the skin temperature.
> - The higher the humidity, the less efficient evaporation is; body cooling via sweating is less effective.
> - High humidity makes the heat more uncomfortable.

body cooling via sweating is less effective. High humidity makes the heat more uncomfortable and everybody, active as well as inactive, is likely to feel uncomfortably sticky out of an air-conditioned environment. Air movement can increase both heat and evaporative loss. Radiant heat gain through direct sunlight or from surfaces such as that of a tennis court can also contribute to heat stress.

In an attempt to provide meaningful information about 'hot' weather conditions and the degree of heat stress for the exerciser it is possible to refer to a heat stress index such as the Wet Bulb Globe Temperature (WBGT) Index. The WBGT uses the following measures:

1 humidity (measured with a wet bulb thermometer);
2 temperature as affected by radiant heat from the sun (measured with a black globe thermometer); and
3 air temperature (measured with a dry bulb thermometer).

The formula is as follows:

$$\text{WBGT } (^\circ\text{C}) = (0.7 \times \text{wb}) + (0.2 \times \text{g}) + (0.1 \times \text{db})$$

(where wb = wet bulb temperature, g = black globe temperature, db = dry bulb temperature)

So, for example, if wb = 26°C, g = 45°C, and db = 32°C, then

$$\text{WBGT} = (26 \times 0.7) + (45 \times 0.2) + (32 \times 0.1) = 30.4^\circ\text{C}.$$

Clearly, the wet bulb temperature has the greatest influence on the WBGT as it accounts for 70 per cent of the index. This highlights the fact that it is the humidity that has the greatest impact on heat stress.

The WBGT guidelines are applied to the exerciser who is lightly clothed. In the WBGT range 26.5–28.8°C, it has been recommended that the untrained, and those who have not undergone heat acclimatization, should be prudent; in the 29.5–30.5°C range, it is recommended that high-intensity exercise be avoided; above 31.2°C it is advised that training should not take place. While the above are recommendations, elite athletes may have little choice but to compete, despite the fact that the WBGT is above 32°C. It is very unlikely that events in prestigious championships will be cancelled because it is too hot and humid.

Fig 5. The exerciser in heat should make an assessment of the prevailing weather conditions before training or competing.

(Reproduced with permission from Mr Alan Ramage, Law Amateur Athletic Club)

CONFLICTING OPINIONS FROM RESEARCH

The messages from laboratory studies often seem to contradict findings made 'in the field'. This may simply be because, though the situations may seem the same there may in fact be differences in the tests, for example, in some laboratory studies athletes are asked to maintain a given speed, whereas in actual events the athletes vary speeds, and are likely to slow down in very hot conditions. Noakes (2001) highlights that trials at a given intensity are less reproducible than when the athletes are able to select their own intensity. In a 'hot' laboratory, the only heat loss mechanisms are evaporation plus some pendulum convective heat loss through limb movement. By contrast in the 'hot' field, in addition to the above heat loss routes, there is air movement over the whole body. This convective cooling effect is more marked in cycling compared with running, due to the higher speeds over the ground. Increased air movement over the body would be conducive to increased evaporation of sweat. The limited convective cooling in the laboratory may result in a higher sweat rate than in the field and produce values that may not occur in competition. Thus, there may be greater levels of dehydration and more emphasis placed on the need for fluid intake as a result of these laboratory studies. An anecdotal observation by one of the authors is that after exercise involving running or cycling, in cold and heat (35°C), liquid sweat has only appeared on the skin after stopping, especially if going straight indoors. This illustrates the importance of convective heat loss, and is almost the equivalent of going from the 'field' into a 'laboratory'. It also has possible implications for the onset of heat stroke when exercise movement stops. There may be an overestimate of dehydration in the field. The runners may start hyperhydrated (having drunk an extra quantity of water), or there may have been limitations in the accuracy of the weighing. The results from the laboratory may not be totally applicable to the field, and vice versa.

CIRCULATORY STRAIN DURING EXERCISE

During exercise in heat there is a demand for blood to carry oxygen (O_2) to the muscles, and to transport waste products from the muscles. In addition, there is a need for an enhanced skin blood flow to promote heat loss. There is a limited amount of blood available to carry out these roles in heat. The upright human may encounter problems in heat when the redistribution of blood results in a large volume of blood collecting in veins below the 'pump' (heart), known as blood pooling. This blood pooling reduces the filling pressures in the heart, and may result in a lower stroke volume and cardiac output. A reduction in stroke volume (the volume of blood ejected by each beat of the heart) necessitates an increase in heart rate to maintain cardiac output (volume of blood ejected by the heart per minute). In exercise of a very high intensity it may not be possible to increase heart rate sufficiently to make an effective increase in cardiac output.

The cumulative effect of sweat loss during exercise in heat often results in a decrease in blood volume and a fall in cardiac output. The competing needs of the working muscles and the skin create a simultaneous demand for blood from the working muscles and skin. This situation creates the potential for a diminished regulation of the heat load.

Blood pooling can promote short-term circulatory failure resulting in fainting (syncope). This happens most frequently in the heat and can occur when a person stands still for a long period of time (*see* page 43), but it can also happen immediately after a person stops running. The common factor is the absence of muscle contractions pumping the blood back to the heart.

PERFORMANCE

There are many reports of endurance athletes who have not performed well in heat and anecdotal comments are supported by laboratory evidence. Galloway and Maughan (1997) showed that endurance time, at a level of effort requiring 70 per cent of their maximal oxygen uptake (VO_2max), was much lower at an ambient temperature of 31°C compared with 11°C. At 31°C the mean endurance time was 52 minutes compared with 94 minutes at 11°C. Possible causes for decreased endurance performance in heat include a high core temperature, dehydration, and a higher perception of effort. Heat production may be so great that the body is unable to dissipate sufficient heat to avoid a marked increase in core temperature, so that hyperthermia becomes a potential problem. Gonzalez-Alonso et al. (1999) reported that high core temperatures were a cause of fatigue in endurance cycling in hot conditions. They discussed the findings of Nielsen et al (1993), who have shown, in a number of laboratory exercise tests in the heat, that their subjects stopped at a core temperature of about 39.7°C on a regular basis. The subjects had improved their exercise time from 48 to 80 minutes as a result of heat acclimation sessions over a 9- to 12-day period but still stopped exercising at around the same core temperature, despite running for much longer after acclimation. Possible explanations for a high core temperature being associated with fatigue include a reduction in the recruitment of skeletal muscles and a decreased motivation. It would appear the brain stops the exerciser to prevent heatstroke. Some athletes may drive themselves beyond safe limits but it is not clear why they are able to override this regulation.

Marino et al. (2004) showed that East African endurance athletes performed better in distance events in the heat than Caucasian runners, but that there was no difference between the groups in cool conditions. They suggested that, because of their small muscle mass and body size, they have a higher ratio of surface area to body mass. This will give them more surface area available for heat loss per unit of body weight than larger heavier athletes. At a given speed, a lower muscle mass results in reduced heat production compared with a larger muscle mass. Thus, in hot conditions, small runners are less likely to increase core temperature to a critical level, which would make them run at a reduced speed. Marino et al. (2004) suggested that Caucasian runners decreased running speed in the heat to prevent the development of hyperthermia.

HEAT PRODUCTION

During exercise there can be a marked increase in heat production, which, if sustained in hot conditions, can result in thermoregulatory problems for the exerciser. Exercise results in an increase in heat production, as the working skeletal muscles are heat generators. Around 75 per cent of the energy produced by the skeletal muscles is given off as heat. For the athlete who exercises in heat, the main means of heat loss is via evaporation. A considerable heat load may develop, particularly in hot and humid conditions, when evaporative heat loss is limited.

Resting oxygen uptake (VO_2) is around 250ml/min, and heat production is 1.2kcal/min. Elite endurance runners may be able to sustain a running speed that requires a VO_2 of 4ltr per minute, with heat production in the region of 20kcal/min. Thus, when distance events are held in hot conditions, there is the potential for the development of heat-related problems. This is particularly true for athletes who can run very fast for relatively short periods of time.

The following gives an example of heat production and the evaporation of sweat necessary to prevent heat gain for an endurance runner who maintains a speed that requires a VO_2 of 4ltr per minute for a period of 1 hour:

1ltr VO_2 = 5kcals

VO_2 of 4ltr per min = 20kcals

20kcals per min for 60 mins = 1200kcals

75 per cent heat production = 900kcals

Evaporation of 1ltr of sweat results in 580kcals heat loss.

900/580 = 1.55ltr

900kcals needs 1.55ltr of sweat per hour to prevent heat gain.

In fact, it is likely that a sweat rate of closer to 2ltr per hour would be necessary, as some sweat will roll off the skin. Since evaporation of sweat is necessary for heat loss, sweat that rolls off the skin does not contribute to heat loss.

Increasing the speed or intensity of the exercise results in a higher heat production, sweat rate and core temperature. Maintaining a high running speed, for example, will result in greater heat production compared with lower speeds. Clearly, runners and cyclists will want to maintain the highest speed possible in an attempt to win their race, but it may be wise to consider tactical alterations in hot conditions. It is acknowledged that exercise intensity can vary in some sports. For example, in cycling, tactical strategies, changing gradients and the amount of shielding will all influence the intensity of the exercise and, thus, heat production. In many team sports, such as soccer, exercise intensity may not be constant. One way of avoiding sustained high-intensity exercise in hot conditions is to change the team tactics to emphasize the passing game.

The main determinant of core temperature during exercise in heat is exercise intensity through a wide range of environmental conditions. The magnitude of core temperature increase is fairly independent of the environment. However, the combination of a high metabolic and environmental heat stress can result in an increased core temperature. Environmental heat stress will add to body heat gain when the air temperature is above the skin temperature. In high humidity, evaporative cooling may not be sufficient to dissipate heat.

DEHYDRATION

Sweat

Even during moderately intense exercise it has been shown that heart rate and sweat rates are higher when the ambient temperature is raised from 22°C to 35°C, while humidity levels remain similar. When the ambient temperature is held at 35°C, and the humidity is increased from 50 to 90 per cent, sweat rate and heart rate are increased further. Clearly, the evaporation of sweat is less efficient in higher humidity, and there is an added circulatory strain.

Fox et al. (1966) quote the following example:

> sweat rate at an ambient temperature of 35°C and humidity of 50 per cent = 1.0ltr/hour; heart rate 150 beats/min
>
> sweat rate at an ambient temperature of 35°C and humidity of 90 per cent = 1.6ltr/hour; heart rate 165 beats/min
>
> (From Fox et al. (1966)

Examination of sweat volumes and the components of sweat may provide guidance on fluid intake for the exerciser. In a group of exercisers who are exercising at the same intensity, and in the same environmental conditions, sweat rates and sweat composition can vary greatly between individuals. These variables can also differ markedly for a given individual over time, influenced by his training status and level of heat acclimatization. An individual who enhances his aerobic fitness and carries out heat acclimatization will increase his sweat rate for a given intensity; there will also be a decrease in the sodium content of his sweat.

Water is a major component of sweat, but it also contains a number of electrolytes and organic components. It contains sodium and chloride, but the ion concentrations are normally lower than those found in the plasma. Potassium is lost in sweat, but plasma potassium levels may be increased from a number of sources including the liver, skeletal muscles, and possibly as a result of red cell damage during exercise (Maughan, 1991). The above has implications for the selection of the optimal composition of fluid for the exerciser in heat. For example, if the concentration of sodium chloride (salt) is higher in the plasma than in the sweat, it seems reasonable to question whether the replacement of salt is of critical importance during events in hot conditions (*see also* page 50).

Fluid Loss

The evaporation of sweat is essential to minimize the increase in core temperature during exercise in hot conditions. While the evaporation of sweat facilitates cooling, it also promotes fluid loss, which can have a detrimental effect on the circulation, body-cooling capability, and exercise performance. An added source of fluid loss is through the respiratory tract, particularly when the athlete is exercising hard in hot dry conditions when large volumes of air are inspired and expired. These conditions are often found in desert climates, and at altitude.

The volume of sweat loss during high-intensity exercise may be as much as 2ltr/hour. It is unlikely that the exerciser will be able to consume fluids at this rate to compensate. Fluid losses of this magnitude may be essential to limit the rise in core temperature via evaporative sweat loss, but such sweat rates will lead to progressive dehydration and a loss of electrolytes.

The following are some of the possible adverse effects of dehydration:

- decreased endurance performance;
- loss of heat acclimatization benefits;
- increased submaximal heart rate;
- decrease in stroke volume;
- impairment of thermoregulation;
- increase in core temperature;
- decrease in the ability to tolerate heat;
- increased chance of heat illness;
- increase in perception of effort.

To study the effects of dehydration Armstrong et al. (1985) used a diuretic (which promotes urine production) to dehydrate their subjects by ~2 per cent of body weight. The 5k times for hydrated and dehydrated (1.6 per cent of body weight) were 18.2 and 19.5 minutes respectively. The 10k times for hydrated and dehydrated (2.1 per cent of body weight) were 38.9 and 41.5 minutes respectively. The above research shows that large decrements in performance occur when subjects are dehydrated. Since races are often won in less than half a second, these data suggest that, purely on the basis of performance, the exerciser should avoid dehydration in endurance events.

However results from an Ironman triathlon, where the athletes were divided into groups based on the weight loss during the triathlon, suggest that large weight losses may not have a great influence on endurance performance. Despite one group having an average 6 per cent (range 5.0–10.7 per cent) body weight loss (due to dehydration) compared with –3.9 and –2.1 per cent weight loss in the other two groups, endurance performance was not related to decreases in body weight (Sharwood et al. 2004). Athletes who had large weight losses apparently did not have a marked reduction in performance. In fact, those losing most weight tended to have the fastest times.

According to some laboratory studies the 6–10 per cent body weight decrease through dehydration reported by Sharwood et al. (2004) would be expected to result in 58 to 125 per cent decrease in performance. It should be highlighted, however, that Sharwood et al. (2004) described the race as taking place in 'mild environmental conditions'. They contend that some laboratory studies do not reflect outdoor exercise, which lasts for many hours. Some reasons for possible differences in laboratory and field findings have been outlined above and are expanded here.

Some laboratory studies have induced dehydration by using a diuretic, which removes the fluid from the intravascular space (reduces the volume of circulating blood), leaving the subject with a relative hypovolaemia (lack of volume in the circulation) at the start of the exercise. When the exercise starts, because of the reduced volume, the heart has to work harder than normal to provide circulation to the working muscles, and to the skin, to allow for the increased blood flow needed for heat loss and for fluid to the sweat glands. The reduced circulating volume could result in impairment of the mechanism for removing the excess heat from the cells, and also the mechanism for heat loss from the skin by having insufficient fluid to fill all the available skin capillaries, and insufficient fluid to allow the sweat glands to function at their best. This would mean that the core temperature would tend to rise faster than it would with a full circulating volume.

In the field situation the athlete normally starts with a full circulating volume. As fluid is gradually lost from the circulation through sweating, the exercise itself causes fluid transfer, first from the interstitial space (the space between the cells, in other words, outside the circulation and outside the cells) into the intravascular space (the circulation) to replace the fluid loss (Tappan et al. 1984). If fluid loss continues, the exercise causes fluid to be transferred from inside the cells to the interstitial space (Tappan et al., 1984). Thus the endurance runner would finish the event with some loss of fluid from the interstitial space, and possible slight dehydration of the cells, but with limited depletion of the circulating fluid volume. This loss would mean that the mechanism for removing the excess heat from the cells would not be too impaired. The heart would also be under less stress.

Dehydration, and the consequent reduction in body weight, would decrease the oxygen cost of the activity. However, Coyle (2004) cautions that this interpretation may not be appropriate as there are no studies to support it. He emphasizes that the combination of dehydration and hyperthermia is likely to reduce cardiovascular function, and that any benefit would only be realized if the athlete could combat the hyperthermia resulting from the dehydration. Coyle (2004) considers that it is unusual for dehydrated athletes not to have an increased core temperature compared with a euhydrated state.

Given that dehydration and hyperthermia combined will synergistically promote fatigue, it is difficult to make a case for voluntary dehydration to reduce body weight and lower energy costs. A strategy that results in the early attainment of a fatiguing core temperature as a result of voluntary dehydration is not sound. It is likely that fast runners will find it difficult to take fluid on board at high speed, and, along with the limited gastric emptying (due to high exercise intensity), this could result in a marked level of dehydration. Coyle (2004) concludes that there is reasonable consensus that dehydration should not exceed 2 per cent in most athletic contests. However, Noakes (2001) counsels that more research is needed on the effects of dehydration. He stresses that studies should focus on dehydration in a long-duration competitive setting, and that the statement that athletes should not lose more than 2 per cent body weight during exercise should be assessed. In his review, Coyle (2004) states that the magnitude of the effect of dehydration on endurance performance will depend on environmental conditions. Ambient temperatures of about 20°C with dehydration of 1–2 per cent body weight have resulted in little effect on endurance performance under 90 minutes, whereas in events over 90 minutes endurance performance has been reduced. Dehydration of 2 per cent body weight in ambient temperatures of about 30°C has led to a reduced performance in endurance activities of longer than 60 minutes. This suggests that athletes will cope less well with dehydration of 2 per cent body weight in hot conditions than in temperate conditions and consequently should avoid dehydration in hot environments when possible. According to Coyle (2004), it is important to identify the activities, individuals and environmental conditions in which dehydration of around 1–2 per cent is not likely to contribute to heat illness and may enhance performance.

It has become accepted that dehydration reduces skin blood flow and therefore also reduces sweating thus increasing the risk of heatstroke. However Noakes & Speedy (2006) cite evidence that sweating is regulated independently of skin blood flow but requires an intact neural supply to supply the sweat glands. The absence of sweating in some cases of heatstroke is usually attributed to dehydration but alternatively it may be due to the high core temperature affecting the neural control of sweating, as well as other cerebral functions. Cade et al. (1971) proposed that fluid loss from sweating during exercise causes a loss in blood volume, which adversely affects the ability of the body to dissipate heat, and may, indeed, be of major importance in the genesis of heat stroke. However, despite one group having an average 6 per cent (range 5.0–10.7 per cent) body weight loss (due to dehydration) after an Ironman triathlon, increases in core temperature were not related to decreases in body weight as compared with others who lost less weight (Sharwood et al. 2004). Athletes did not experience exceptionally high core temperatures or an increased risk for medical treatment as a result of heat illness. Noakes (2001) stresses that a number of studies using field trials have found no meaningful association between the degree of dehydration, and the end-of-race core temperature.

Heatstroke appears to be common when vigorous exercise is undertaken with inappropriate clothing, for example as happened with an overdressed man shovelling snow and with military personnel who had to run across mudflats on a hot day wearing wetsuits. Noakes (2001) highlights the fact that runners are at most danger of heat injury in races that last from 15 to 60 minutes (between 5 and 21k for elite performers, who run at ~85 per cent of their VO_2max, and above). Alberto Salazaar was a one-time fastest male marathoner who, after running very fast in a 7-mile race in very hot conditions, was covered in ice and given the last rights. However, he did recover and was able to compete again.

The main determinant of core temperature during exercise in heat is exercise intensity, and the combination of a high metabolic and environmental heat stress can result in an increased core temperature. This would inevitably mean that fast finishers had higher core temperatures, which fits in with Coyle's (2004) observation that despite running in cool conditions, runners who were dehydrated (often fast finishers) sometimes had core temperatures in the region of 40°C.

A rise in core temperature will result in an increase in the temperature of the blood circulating through the skin. This in turn will increase the temperature differential between the skin and the ambient air, and increase the rate of heat loss. This will have a marginal effect in a laboratory but may have more effect in the field because of convective heat loss resulting from the athlete's movement over the ground. This assumes that the ambient air temperature is lower than the skin temperature.

STRATEGIES FOR EXERCISING IN THE HEAT

All athletes who live in a temperate climate should recognize the difficulties that may arise when they live and exercise in heat. While endurance athletes are at greatest risk of performance decrement and health problems while living and training in the heat, other athletes should not be complacent. The athlete should be aware that simply living in hot conditions can cause chronic dehydration. Sprint and short-distance endurance athletes are often involved in heats and several warm-ups. They are regularly exposed to heat when training and competing in many foreign locations, and sometimes at home. Games players normally play for between 80 and 120 minutes (including extra time), and often have a warm-up of 40 to 50 minutes. Dehydration and carbohydrate depletion are possible causes of reduced performance.

There are a number of strategies to minimize the effects of playing in heat, to attempt to counter the adverse effects of dehydration and carbohydrate depletion (*see also* Chapter 3).

Prior Experience

Athletes who attempt to compete in a major event when they have not experienced living and performing in heat are at a potential disadvantage. It is prudent to experience living, training and, if possible, competing in heat before a major event. Regardless of the sport, prior experience of heat will help athletes determine if they are particularly susceptible to heat, alert them to specific problems, and help them identify the efficacy of their coping strategies. The same applies to living in a hot environment. While general guidelines have been formulated for fluid intake and heat acclimatization in a hot environment, they are not likely to be appropriate for all. Experience of heat will allow the athlete to experiment, in order to help counter any potential difficulties. It will also highlight the need for possible adjustment to methods and procedures to minimize the effects of heat. Early experimental work with fluid intake, monitoring of body fluid status, and manipulation of heat acclimatization strategies (Chapter 4) are all worthwhile.

Experience of a hot environment can be revealing. The athlete may prove to be a 'heavy sweater' who will need to have a very high fluid intake. Some athletes may perform particularly badly in heat, and need a longer heat acclimatization period. An individual who plays a game that is characterized by lots of running, and who does not cope well with heat, may need to be substituted. Alternatively, the tactics of the individual or the team may need to be amended to allow for the hot conditions.

Adjustment

Adapting to the lifestyle in a hot environment is important. For example, it will be necessary to devise strategies to ensure that it is easy to maintain regular fluid intake. A change of location usually disrupts the normal food and drinking patterns of the athlete, but travel from a temperate to a hot climate will always increase the need for fluids. The athlete should not only regulate fluid intake to help cope with the hot conditions, he must also establish a routine to facilitate the appropriate intake.

It is easy for an athlete to become dehydrated, particularly during a long flight in an aeroplane cabin with very dry air. There may be other reasons why he will arrive badly dehydrated: disruption to routine, problems with accessing fluid, lethargy after the long flight, and the much-increased demand for fluid in a hot climate. While the lethargy may be partly caused by jet lag, dehydration could be a contributory factor.

Avoidance of unnecessary exposure to direct sunlight will reduce sweating and lessen the possibility of dehydration. Athletes should wear loose, light-coloured clothing (*see* below) when they move around (for example, during sightseeing), and avoid sunbathing. They should seek shade whenever possible.

Exercisers should be encouraged to drink while they are living in a hot environment. A sedentary individual in a cool environment is likely to need about 3ltr of water per day, but this will increase to 3–5ltr per day if he exercises for just an hour. In hot conditions the demand for water will increase to 4ltr per day for the sedentary individual but, if the individual exercises for two hours per day, it may be as high as 10–15ltr per day. Such volumes might be quite difficult to consume. (Incidentally, care should be taken when large volumes of fluid are ingested. The energy content of the fluids may be quite high, and could result in weight problems, particularly for weight category athletes.)

Heat acclimatization benefits are lost if the exerciser is dehydrated, as are the benefits that an aerobically fit individual has compared with an individual in an unfit condition (*see* Chapter 4).

Warm-Up

Athletes fatigue when a given core temperature is reached. A lower pre-exercise core temperature has the potential to enhance endurance performance. A common aim of a warm-up is to increase muscle and core temperatures. In heat this strategy may be counter-productive as the exerciser will start with a higher core temperature, reach the 'slow down/stop' temperature sooner, and be closer to other 'heat problems' than without a warm-up. However, a warm-up may also include a rehearsal of the skills of the activity, for example, in hurdling or soccer, as well as an appreciation of the prevailing environmental conditions, including the playing surface.

Despite limited evidence that a warm-up reduces injury, some athletes also feel that they cannot compete properly without their regular warm-up routine, and that a restricted warm-up will actually increase their chances of injury. There may still be a case for a reduction in the warm-up, however. For example, in a long-distance race, a limited warm-up may not have a major physiological or mental effect. In team sports, consideration should be given to a reduction of 'run-throughs', which should be carried out in a cool shady area. If possible, any warm-up should not be done in direct sunlight. A warm-up in the shade and away from any radiative heat gain from surfaces is desirable in heat. A reduced warm-up (intensity and/or duration) should be practised and perfected long before competition.

Pre-Cooling

If the athlete warms up before competition in hot conditions, he is likely to reach the 'fatigue' core temperature sooner. If it were possible to delay the time that it took to reach this 'failure' temperature, it may be possible to enhance performance. If pre-cooling means that it takes longer before the athlete reaches the 'failure' core temperature during exercise, that athlete should have an improved endurance performance. A lowered skin temperature and core temperature as a result of cooling will reduce sweat rates and skin blood flow. The lowered skin temperature will increase the temperature gradient between the body and the skin, and promote heat transfer to the skin.

A number of strategies have been used to pre-cool the body before exercise in heat (Marino, 2002). Methods of pre-cooling have included exposure to low air temperatures, immersion in water with a gradual decrease in water temperatures, the wrapping of cold, wet towels around the body, lowering the water temperature of a shower over a time period, and the wearing of an ice vest. This variety in cooling methods, along with different time periods for cooling, and a wide range of environmental temperatures, exercise time and mode make a clear-cut evaluation of the effectiveness of pre-cooling difficult.

Studies have reported conflicting results, with some but not all showing that pre-cooling can enhance endurance and performance in short, high-intensity exercise. While pre-cooling (using exposure to cold air or immersion in cold water) can improve performance and inhibit increases in physiological variables such as core temperature and skin temperature, the above strategies have limitations since some procedures involves a long timescale and disruption to warm-up and pre-event behaviours. Cooling may promote shivering, which is unpleasant, and may interfere with skill parameters. Thus the applicability of some laboratory findings to competitive events is questionable. In addition, some pre-cooling studies have failed to show an improved performance. However, the use of an ice vest may enhance performance and cause limited upset to the warm-up procedures and pre-competition rituals.

The use of a cooling vest (with eight pockets for ice packs of about 500ml each) may be of value, as a warm-up can be carried out without major disruption and provide benefit for the athlete in hot environments. In one study (Sigurbjorn et al., 2004) male and female runners who wore an ice-filled vest during their warm-up had an enhanced 5k time (by 13 seconds) compared with a warm-up in a T-shirt. The trials were conducted in a temperature of 32°C and 50 per cent humidity and the ice vest inhibited the increase in core temperature, heart rate, skin temperature and thermal discomfort compared with a T-shirt during the warm-up and early part of the 5k run. These advantages had all but disappeared by 3.2k into the run but the runners had a faster pace in the later two-thirds of the run. The decreases in core temperature and skin temperature as a result of wearing the ice vest were small, but still the athletes improved their performance. The ice vest weighed about 4.5kg and consequently increased the oxygen cost for a given running speed. The authors attempted to minimize this problem by asking the runners to reduce their warm-up running speed. Although a few runners preferred wearing the T-shirt, citing the weight of the vest as a source of discomfort, most runners preferred the ice vest, feeling that they sweated less and were cooler with the vest.

Sigurbjorn et al. (2004) examined performance over a 5k run with times between 15.6 and 22.8 minutes. They refer to the literature when they speculate that pre-cooling may enhance performance in running events between 600m and 10k. They postulate that pre-cooling may not be effective in events much longer than the 10k, as the pre-cooling benefits are likely to be short-term. However, they do concede that there are no data for long-duration events that can provide meaningful information.

Benefits of wearing an ice vest

- The use of a cooling vest (with eight pockets for ice packs of about 500ml each) may be of value, as it allows a warm-up to be carried out without major disruption and provides benefit for the athlete in hot environments.
- Wearing an ice jacket during the warm-up may enhance performance in running events of between 600m and 10k.

Adapting Intensity

Reducing intensity of exercise during competitive events may of course diminish the likelihood of success, but it may prove to be fruitful. In team games, the players may decide to emphasize a passing game. The role of substitutes may take on a new meaning in heat. In hot conditions it may be prudent to substitute a 'heavy sweater', or a player who has a predominantly running role. In a distance race, the athlete may intentionally reduce his pace in the early stages. It appears that in professional soccer players the core temperature rises till half-time but remains steady during the second half.

Increased aerobic fitness allows the exerciser to work at a lower relative intensity and reduced core temperature, or at a higher absolute work rate at a given temperature.

Avoiding Unnecessary Heat Gain

If possible, avoid direct sunlight. Avoid 'hot spots' where possible – substitutes should not sit in the sun, and they should also keep away from synthetic surfaces that may produce radiant heat gain. Athletes should run or cycle in the shade in a race whenever possible.

Sweat-inhibiting deodorants or sunscreens should not be used.

In most competitive events, the rules insist on some form of uniform. Thus, in many instances the athlete will not have any choice of clothing, being obliged to wear, for example, the same soccer shirt, shorts and stockings as the other players. In the few instances where the choice of clothing is negotiable, the athlete should examine the characteristics of clothing when exercising in heat.

Clothing can limit heat loss from the body by providing a layer, or layers, of insulation and consequently limiting heat transfer. In hot conditions, evaporation of sweat is the main means of cooling. Clothing will limit evaporative heat loss from the skin, diminish convective cooling and is therefore likely to increase skin and core temperatures. Evaporative sweat rates are reduced when the amount of clothing and the number of layers are increased. Factors that influence evaporative heat loss from the skin include wind speed and the vapour pressure gradients relating to the skin, clothing and ambient air. Exercise in heat increases sweat rate and decreases the vapour gradients, leading to reduced heat loss and greater wetness of the skin. An increased wetness on the skin is perceived as being uncomfortable. Wearing clothing is likely to increase the wetness of the skin, and the perception of discomfort.

A number of manufacturers have developed clothing that promotes evaporation by improving transfer of fluid. Despite the introduction of such materials, there is limited evidence that shows that their characteristics make a meaningful difference to the athlete; admittedly, there appear to be few trials in this area. One study by Gavin et al. (2001) compared three different conditions – subjects semi-nude, wearing cotton, or wearing an evaporative polyester fabric – in 30°C and 35 per cent humidity during rest and running at 70 per cent of their VO_2max. They concluded that wearing an evaporative polyester fabric did not provide any apparent advantage. There were no meaningful differences among the conditions for skin and core temperature, heart rate or thermal comfort, despite the fact the polyester fabric did enhance evaporation.

Some soccer players wear a vest and a shirt, even in hot conditions, although there seems to be no sound scientific evidence to support this practice. It is possible that the vest will wick fluid and enhance evaporative cooling, but one study in our laboratory failed to show any advantage of a vest and shirt against a shirt. In fact, the best condition was semi-nude, when the subjects wore only shorts. In the semi-nude condition, endurance time was longer than with a shirt or shirt and vest.

Considerations for the choice of clothing for exercise in heat include the impact of clothing on performance, and how the clothing feels. It has been suggested that 100 per cent cotton may not 'feel so good' in very hot conditions compared with a mixture of fabrics. When the athlete exercises in heat, it is prudent to wear loose-fitting clothing to promote air movement, as long as aerodynamic factors are not important. Studies have shown that a nude surface has a greater evaporative heat loss than a clothed surface, but competition rules in sports dictate that some form of clothing is necessary. A large vapour gradient between the skin, the clothing and ambient air is desirable to promote evaporative heat loss. Thus, clothing that facilitates the removal of moisture from the skin may reduce the vapour pressure on the skin and promote evaporative heat loss.

If the athlete is likely to encounter heat problems, the following should be considered:

- Dark-coloured clothing should be avoided in sunny conditions, as it will absorb heat.
- If possible, the athlete should wear porous clothing to facilitate evaporative heat loss.
- Any type of clothing that will restrict evaporative heat loss should be avoided.
- If a layer is worn, it would make sense for it to be loose-fitting.
- If wind resistance is not a consideration, loose-fitting clothing should be worn, to enhance convective heat loss.
- In direct sun, a sun hat should be considered, but a hat will limit evaporative heat loss.

Tennis players sometimes change a wet shirt for a dry shirt during a match, either to improve thermal comfort or reduce weight or drag on the skin. However, a shirt needs to be wet to promote evaporation of sweat, so the change to a dry shirt may be counterproductive.

Fatigue in Heat

The following factors may be possible causes of a decrement in exercise performance in heat:

- hyperthermia (high core temperature); it has been shown by some researchers that exercisers will stop exercising when their core temperature reaches a certain level;
- dehydration (reduced fluid levels in the body) can affect performance particularly in a hot environment;
- hypoglycaemia (low blood-sugar levels) can cause disorientation and reduced work capacity;

- glycogen depletion (a reduction in liver and skeletal muscle glycogen stores) can lead to a reduction in endurance performance;
- hyponatraemia (low blood-sodium levels) causes a general malaise and fatigue, often with bloating of the body, and can be life-threatening. Normally associated with long-duration events (4 hours and above) during which large volumes of fluid are consumed), resulting in 'water intoxication', and a reduction in blood-sodium levels;

Fig 6. A sunhat may prevent radiant heat gain during exercise in the sun but is likely to limit evaporative heat loss.

(Reproduced with permission from Charles Campbell, aka Charlie Campbell; *see also* www.munrorun2000.co.uk/gallery)

Strategies to combat heat/humidity

- Assess prevailing weather conditions.
- Recognize the environmental factors likely to cause heat disorders.
- Identify the possible causes of fatigue in heat and devise strategies to counter them.
- Gain prior experience of living and exercising in heat before important competitions in heat.
- Acclimatize by exercising in heat/humidity for a period prior to competition.
- Identify possible causes of unnecessary heat gain and avoid them.
- Rehydrate after training and competition.
- Assess clothing while exercising, and while living in heat.
- Where there is a choice, choose clothing that is loose-fitting and white.
- Protect the skin.
- Consider adjustment of exercise intensity in heat to minimize heat gain.
- Investigate possible reduction in the extent of the warm-up.
- Experiment with pre-cooling using ice vests.

- salt depletion from sweating causes tiredness, irritability, giddiness, fainting, cramps and loss of performance. During the first 7 days in the heat it may be worth adding additional salt to the diet. After heat acclimatization, which takes about 7 days, the salt level in the sweat is reduced (*see* Chapter 5).

Fluid intake using a range of solutions has the potential to reduce the adverse effects of some of the above causes of fatigue. Fatigue may result from one of the above factors (or others), or from a combination of factors. There are instances where a strategy to enhance performance will not have any impact on other possible causes of fatigue. For example, large concentrations of carbohydrate may limit gastric emptying and delay fluid reaching the cells, and will not help counter the adverse effects of dehydration. Carbohydrate will not prevent a rise in core temperature.

Identification of the possible causes of fatigue will help the athlete formulate an appropriate strategy to minimize the problem.

There are a number of preparatory strategies that an athlete may adopt in order to promote performance in heat:

- Have trials of lifestyle, training and competition in heat to identify problems.
- Have a trial of strategies to minimize heat problems: restricted warm-up, fluid intake, monitoring of hydration status.
- Assess 'heat' lifestyle, including fluid intake before/during/after exercise, fluid availability during day, at meal times and at training, and in the bedroom.
- Check effectiveness of fluid intake by monitoring body weight and urine colour.
- Identify whether an athlete has particular heat problems.
- Check appropriateness of clothing.

After Exercise

Showers following exercise should not be too cold because cold on the skin stimulates heat production and causes vasoconstriction, thus reducing heat loss. Anecdotally, several people have found that after hard exercise in the heat, resulting in profuse sweating, a cold shower was pleasant, but sweating started again immediately after coming out of the shower. A good cooling method is to allow a cool shower to play only over the head and neck, areas that can lose a lot of heat without vasoconstriction. This preserves the heat loss system over the rest of the body, is very comfortable, and protects the brain from the danger of overheating. The hands and wrists and the feet and ankles are regions with a large surface area to body mass ratio, and cool water over these areas will help accelerate cooling. Fanning the face is another effective way of starting the cooling process. On the field, towels kept in cold water can be placed over the head and/or draped round the neck, or iced water can be used to immerse hands and arms.

Air Conditioning

In a consistently hot climate it is likely that most, if not all, buildings will be air-conditioned. This raises its own problems because of the marked difference between the outdoor and indoor temperatures; at the extreme, it may involve moving between temperatures of 12°C and 39°C. Going out will feel like walking into an oven while coming back into the air-conditioned environment will produce a marked feeling of chill, and may lead to shivering, even with the air conditioning kept at 21°C. (The body responds to *change* of temperature rather than to absolute temperature.)

The repeated chilling effect may result in an increased risk of developing colds and other respiratory infections. Those who feel the 'chill' effect of air conditioning should put on more clothing when they come indoors, and make a concerted effort to keep away from anyone who has a cold. Anybody who is sweating when he comes in from the outside heat should dry himself and change into dry, warmer clothing if he is going to be in the air conditioning for more than a few minutes. It has been noticed that upper respiratory infections are related to frequent, rapid changes in temperature. These are also associated with increased respiratory secretions (Lloyd, 1986). If the athlete has been exercising hard in the heat, much of the breathing will have been through the mouth, which has the effect of drying the airways. The body will respond by increasing fluid secretion. Coming into air conditioning will remove the drying due to mouth breathing and to heat, but the secretions are likely to continue for some time, producing a very moist respiratory tract. This creates an ideal culture medium for any viruses that may be going around.

Sleep

High bedroom temperatures are associated with a greater incidence of disruption to sleep, and a range of 16–23°C has been suggested to promote good sleep. A water bottle by the bedside can also enhance fluid intake.

DEHYDRATION AND HEALTH

Dehydration is a particular hazard in a hot dry atmosphere because sweat evaporates quickly, and therefore the person may not be aware of the volume of sweat being lost, and may not even be suffering too severely from the heat.

If sweating is reduced because of dehydration, or because the person has a reduced ability to sweat, heat can be lost by increasing the rate of breathing and by panting. This hyperventilation has been noted in runners just before they collapse with heat injury. Hyperventilation produces some cooling but it does cause loss of fluid and results in a decrease in carbon dioxide (CO_2) levels in the blood. During exercise this falling CO_2 is compensated by the CO_2 produced by the exercise. After stopping exercise, the hyperventilation may eventually lead to unconsciousness and muscle spasm. Rest and fluid replacement (*see* Chapter 4) will promote recovery from the above. Getting the casualty to re-breathe into a paper bag will hasten recovery.

Dehydration results in a reduction in central blood volume. Distressed athletes are likely to respond well as a result of simply being lain down, which results in a restoration of the central blood volume.

Table 1 **Possible adverse effects of dehydration in the heat**

Decreased	*Increased*
Endurance performance	Perception of effort
Ability to tolerate heat	Risk of heat illness
Sweating	Core temperature
Stroke volume	Submaximal heart rate
Loss of heat acclimatization benefits	
Impaired thermoregulation	

HEATSTROKE

Defining Heatstroke

Heatstroke develops when the body is unable to get rid of the excess heat being produced. There will be a varying combination of high environmental temperature, high humidity, lack of wind, vigorous activity, and heat-retaining clothing. Dehydration reduces sweating, thereby increasing the risk of heatstroke, and the effective wind speed depends on which direction the athlete is running. In practical terms it is more useful to refer to a 'heat casualty' rather than to try to differentiate between heat exhaustion, dehydration and heatstroke.

Early symptoms of heatstroke include excessive sweating (which may cease), headache, nausea, dizziness, hyperventilation, and clouding or progressive loss of consciousness, or hallucinations. There may be muscle twitching or convulsions and loss of control of the body sphincters. In severe cases there may be

Diagnosis of heatstroke

- Consider the thermal implications of the clothing and level of activity.
- Patient feels warm/hot.
- Symptoms may include headache, nausea, dizziness, irrational behaviour.
- Sweating may or may not be present.
- Patient may be hyperventilating.
- High core temperature (rectal usually above 41°C).

deep coma, with pinpoint pupils and shock. Elevated heart and respiration rates are often present, with difficulty in breathing, and inhalation of vomit. The patient feels warm or hot and has a high core temperature (rectal usually above 41°C). Sweating may or may not be present. Diagnosis depends on suspecting heatstroke in any hot environment, or if the person is wearing too much clothing for the prevailing ambient temperature, and the level of exercise.

Management of Heatstroke

Without the appropriate treatment a heatstroke victim is in great danger of developing irreversible damage in the brain, kidneys, liver, and adrenal glands, and the end result may be death. Disseminated intravascular coagulation (the blood forming microclots in all the areas of the body and all the organs) may result. This will produce mini strokes and damage to many of the organs. Uncontrollable bleeding may finally occur, because the body has used up all of its clot-forming chemicals.

Treatment must be started *early*. The longer the core temperature is elevated, the more serious the organ damage will be. Unnecessary cooling is much safer than waiting for a definite diagnosis. Body cooling and rehydration are the tactics employed to treat heatstroke and should be carried out as quickly as possible. There is some controversy about the safest way of lowering the core temperature, but the debate can only be resolved by considering the mechanism and pathophysiology that lead to the high core temperature. There are two types of heatstroke:

1. *passive*, when the subject has been exposed to high ambient heat stress for a prolonged time (a number of days). The contribution from muscle-generated heat is *minimal*, but there is likely to be impairment of the heat-dissipating mechanism, which may include an element of dehydration. The temperature rise is slow. This is how heatstroke develops during a heat-wave among people on medication, among the elderly and among devotees on the pilgrimage to Mecca.
2. *active* (exertional heatstroke), when the temperature rise is rapid due to a heavy heat load generated by vigorous muscle activity, often superimposed on a high ambient heat stress, which may include excessive insulation of the body surface during exercise. This is the mechanism in sport, in malignant hyperthermia (an abnormal response to some anaesthetic agents), and during 'raves' following the ingestion of drugs such as amphetamine or ecstasy. As the temperature of cells rises, the metabolic process accelerates.

Treatment

Passive Heatstroke

In the case of passive heatstroke, ice, cold water or cold air on the skin will reduce heat loss by inhibiting sweating and stimulating superficial vasoconstriction. They may also trigger shivering and increased heat production (Khogali, 1983). The combination is potentially disastrous, with an accelerating rise of temperature leading to death. The safest cooling technique is to have the casualty, covered only with a sheet, lying on a string hammock stretcher over a bath. He is sprayed with *tepid* water (15°C) and *warm* air (33°C at the patient) is blown over him via a fan. Vasoconstriction and increased heat production are avoided (Khogali, 1983) and a cooling rate of 0.06°C per minute was reported. Khogali et al. (1983) found, in their study from different centres, that there was a death rate of between 4.7 per cent and 18.4 per cent, but these were among a predominantly elderly and frail population. This has been found to be the safest method of cooling when dealing with the Mecca pilgrims (Khogali, 1983).

Exertional Heatstroke

In exertional heatstroke the main problem is the excess heat production from the muscles. These being immediately under the skin can be cooled by very cold/ice water directly through the skin. There is no problem with vasoconstriction because of the massive heat production immediately under the skin, and any central stimulus for increased heat production is ineffective because the muscles are already in heat production overdrive. Gaffin and Hubbard (2001) and Armstrong (2000) have shown that ice water is the most effective means of lowering core temperature in exertional heatstroke. In their comparison of ice-water immersion (torso and upper legs immersed in water) and air exposure (with wet towels on the torso and part of the limbs), Armstrong et al. (1996) concluded that the former cooled hyperthermic endurance runners (35 ± 3 years) around twice as fast as the air exposure method (0.20°C/min for iced water immersion versus 0.11°C/min for air exposure). There were no fatalities in 252 marine recruits (with rectal temperatures as high as 42.8°C) who were treated using ice-water immersion resulting in a rapid decrease in core temperature.

According to Armstrong et al. (1996), the subject should be placed in cold water (or his torso and upper legs immersed in ice water), or lie on a stretcher above iced water. Sheets are soaked in ice water and placed on the subject and ice should be put on the sheets regularly. Air from a fan can also be directed towards the subject and ice water should be used to cool the head. Gaffin et al. (2000) state that the above procedures normally result in a decrease in core temperature to 39.5°C within 15 to 20 minutes. An added advantage of immersing the subject in water is the fluid shift from the periphery to the central areas, which helps maintain the circulation.

The placing of a heatstroke subject in water may have limitations, including subject discomfort and restricted access to carry out treatment on the subject, and hygiene problems if the subject has sickness and/or diarrhoea. If cardiopulmonary resuscitation is a possible requirement, the subject should be placed on

Heatstroke

- Heatstroke is life-threatening.
- Efforts should be made to decrease core temperature as soon as possible, so that the time and magnitude of the increased core temperature can be minimized.
- Ice-water immersion should be used to treat athletes who have severe hyperthermia or heatstroke resulting from exercise or hard physical activity (Armstrong et al., 1996), while tepid water and cooling fans are used in subjects who have developed heatstroke over a prolonged period or who have known illnesses (Khogali, 1983). The use of ice water for subjects with atherosclerosis is deemed to be inappropriate as the diameter of the arteries in the heart decreases with exposure to ice water.

a stretcher with the ends of the stretcher supported by the side of a bath. This applies to passive (Khogali, 1983) as well as exertional (Gaffin and Hubbard, 2001) heatstroke.

The location, degree of medical support and available facilities will dictate to a certain extent how the heatstroke subject can be treated. In any case, medical help should be sought immediately. The heatstroke subject should be removed from any source of heat (for example, direct sunlight) as soon as possible, and put in the shade, or taken indoors to a cool or air-conditioned room. Treatment must be started *early*, as any delay can cause death. Clothing should be removed. If possible, a rectal thermometer should be used to assess core temperature initially and to monitor the effectiveness of the intervention.

Treatment of heatstroke

- Treatment must be started *early* – delay can cause death.
- Remove from any source of heat into a cool or air-conditioned room.
- Remove clothing within modesty.
- Initiate cooling. Passive heatstroke: spray with tepid water and fan with warm air. If possible, subject should be on a net stretcher over a bath. Exertional heatstroke: place the subject in cold water or lay on a stretcher above iced water and use ice or cold water to cool the subject. Soak sheets in ice water and place on the subject, and put ice on the sheets regularly. Use ice water to cool the head. Direct air from a fan towards the subject.
- If possible, use a rectal thermometer to assess core temperature initially and to monitor the effectiveness of the intervention.
- Insert a urinary catheter to monitor kidney function.
- Rehydrate with sodium-rich fluid such as 0.9 per cent Saline.
- If facilities are limited, it is necessary to improvise. At the very least water can be sprayed on the skin, clothing can be soaked in water, or wet sheets can be placed on the skin and evaporation assisted by fanning.
- Transfer to intensive care unit in hospital.

The individual should be rehydrated with sodium-rich fluid such as 0.9 per cent Saline, with the first 3ltr being given rapidly. A urinary catheter should be inserted to monitor urinary output and, if there is still very little urine production once the patient has been rehydrated, 500ml of 10 per cent Manitol should be given initially to try and restore urinary production.

Cooling should continue until core temperature is below 39°C (102°F), then the patient should be transferred to an intensive care unit in hospital.

At the very least, water can be sprayed on the skin, clothing can be soaked in water, or wet sheets can be placed on the skin, and evaporation assisted by fanning. Cold water can be used to cool the head. If facilities are limited, it is necessary to improvise. A waterproof jacket or polythene could be used to hold water when placed in a hollow in an outdoor setting.

SUNBURN

Sunburn decreases sweat sensitivity and sweat rates, and results in increased body heat storage. Skin protection should be used to avoid sunburn. It has been shown that heat dissipation was not inhibited when exercisers applied sun cream. They performed a 40-minute exercise bout at 55 per cent of VO_2max in an ambient temperature of 32°C and had lower skin temperatures compared with wearing no sun cream.

Causes and Risks

Unprotected exposure to sun is dangerous. Tanning and burning are flip sides of the same coin, and tanning is impossible without risking a burn. Both are caused by ultra-violet (UV) rays, UVA, UVB and UVC. Sunburn can occur unexpectedly and must then be diagnosed and treated.

Unprotected exposure to sun causes sunburn, and depresses immune responses in the skin leading to recurrent herpes labialis (cold sores on lips). It also increases the risk of skin cancer – basal and squamous cell (two types of local skin cancers) on exposed areas, and melanomas (a type of skin cancer that can spread throughout the body) anywhere on the body. Melanomas frequently occur in covered areas on the trunk. Skin cancer is associated with infrequent high doses of UVR and intermittent burning, especially in childhood and adolescence. The risk is greater with exposure, and it does decrease with a sun-protected lifestyle. The risks are also increased by altitude, by reflective surfaces (such as fresh snow, ice, water, sand, metal, concrete), and by wind ('windburn' is exacerbated sunburn). Sunburn can also occur unexpectedly during cloudy weather. Unprotected exposure may also cause skin to age prematurely.

A number of factors increase the risk of sunburn:

- altitude: with every 1,000ft increase there is a 4 per cent rise in UVB;
- reflective surfaces: fresh ice, snow (reflect up to 80 per cent), water, sand, metal, concrete;
- clouds: absorb less than 40 per cent UVR;
- wind: so-called 'windburn' is actually exacerbated sunburn;
- ozone depletion: 1 per cent ozone depletion gives 2 per cent increase in UVR;
- time of day: the greatest sunburn risk is during the middle of the day.

Diagnosis and Treatment

Mild sunburn is shown by erythema (reddening of the skin), warmth, tenderness and pain. Severe sunburn is indicated by the same symptoms, plus blistering.

Sunburn is treated generally by avoiding further exposure, ensuring skin cooling, and considering analgesics and non-steroidal anti-inflammatory drugs.

Mild sunburn can be treated by cool soaking with tap water, by compresses or a bath. The best emollients are those with low concentrations of menthol, camphor or phenol for increased cooling plus anaesthetic effect. Aspirin and other NSAIDs (group of painkillers with some anti-inflammatory action) may be taken, to reduce pain and inflammation. Topical anaesthetics may also be used; pramoxine hydrochloride is best because it is non-sensitizing. Beware benzocaine, which is a photosensitizer.

In a case of severe sunburn (blistering), admission to hospital may be needed for hydration and pain relief. In severe sunburn of the face and neck, difficulties may occur with the airway. Severe sunburn should be treated in the same way as mild sunburn, plus the following: protect bullae (blisters) if intact; sterile drain if necessary; if unroofed or eroded, use topical antiseptic (silver sulphadiazine); monitor for infection; check tetanus status; administer Prednisolone oral (rarely) – 60mg per day tapering in one week – to minimize pain and blistering.

Skin Types

Skin typing is not identical with racial pigmentation. Some blue-eyed blondes tan profusely, while some brown-haired brown-eyed Mediterraneans and some black-skinned people can burn readily.

Table 2 **Genetically determined skin types for sunburn risk and sun protection factor (SPF) for sunburn prevention**

		Minimum SPF to use		
Skin type	*Characteristic*	*Initial exposure*	*Subsequent exposure*	*Never less than*
I	Always burns; never tans	15–50	10–15	8–10
II	Often burns; sometimes tans	10–15	8–10	6–8
III	Sometimes burns; often tans	8–10	4–6	2–4
IV	Rarely burns; always tans	6–8	4–6	2–4
V	Moderately pigmented (brown)			
VI	Heavily pigmented (black)			

(Adapted from Lloyd, E. (1994), ABC Sports Medicine, *British Medical Journal*, 309, 531–4. Reproduced with permission from the BMJ Publishing Group.)

Warnings

1 Tanning parlours use UVA. UVA tan gives less protection than UVB tan and does not prevent sunburn from UVB. Equipment and practices are often dangerous, and melanomas and deaths from burns have occurred.
2 In Caucasians the darkest possible tan is only equivalent to an SPF of 4 and gives minimal protection against burning, skin ageing and cancer.
3 Tan assisters are of no value.
4 The evidence that sunscreen use prevents melanoma or basal-cell cancer in humans is equivocal. Inadequate application of sunscreens may increase the risk because the users may think they have total protection.
5 Quick-tanning agents, oral and surface applied, do not 'tan', but only stain skin.
6 Tan accelerators are not effective.
7 A Psoralen (tan accelerator)–sunscreen combination does accelerate tanning but may also increase the risk of cancer.

Sunburn Prevention

1 Know your own skin type (*see* Table 2).
2 Avoid sun between 10am and 3pm.
3 To block the sun, clothing is better than sunscreens and tight-knit dry fabrics are better than loose-knit wet fabrics. Hats shade the face but allow continued exposure to scattered and reflected UVR.
4 Sunscreens decrease the risk of squamous-cell cancer, skin ageing and possible suppression of immunity. They are measured with regard to sun protection factor (SPF), although this is purely with reference to UVB, and substantivity, in other words, resistance to removal by water, sweat or towelling. Physical screens reflect or scatter UVR, and have high substantivity, but are messy and opaque. Chemical screens absorb UVR and are cosmetically invisible. The best sunscreens currently available contain a combination of protection against UVB (PABA esters such as Padimate O, which have high substantivity, protect against cancer and have a low incidence of sensitization) and against UVA (Avobenzeno or Parsol 1789).
5 There are as yet no sunscreens that can work from inside when taken by mouth or injection.

Applying Sunscreen

Sunscreen use is not an excuse for sun worship. The SPF number is set according to a standard thickness of the screen (2mg/cm^2), and very few people apply enough screen, usually only managing 0.5–1.5mg/cm^2 (Taylor & Diffy, 2002). Ideally, therefore, the highest SPF factor (at least 15) should be used.

One way of trying to apply the correct amount of sunscreen (Taylor & Diffey, 2002) is to adapt the rule of nines, which is used to calculate the extent of injury in the case of burns. The rule of nines divides the body surface into 11 areas each about 9 per cent of the total. If sunscreen is squeezed out on to the whole length of each of the first finger and middle finger, and this amount is spread on to one of the areas of nines, the dose will be approximately correct – the 2mg/cm^2 recommended by manufacturers. One finger will provide half the cover and sunscreen should be reapplied within half an hour. The areas of nines are: head, neck, and face; each arm is one; upper back, lower back, upper front of body, lower front of body are each one; each leg is two.

OTHER PROBLEMS

Prickly Heat

Prickly heat describes a rash of tiny blisters in mildly inflamed skin, which is worst round the waist, upper trunk, armpits and scalp. It is itchy, with intensely irritating prickling sensations. Infection can develop, leading to multiple pimples. The itching prevents sleep till the cooler early hours of the morning and can make the sufferer bad-tempered and inefficient. The cause is unknown but an important factor is the constant wetting of the skin by un-evaporated sweat, which occurs at times of high humidity in hot/wet climates. The prickling can be relieved by taking a cool shower, dab-drying the skin gently to prevent further damage, and then using calamine lotion and zinc oxide dusting powder. Clothing should be starch-free and loose-fitting.

Heat Oedema

The swelling of the ankles occurs during the first few days and settles without treatment. It is similar to the swelling of the ankles after long air flights.

Heat Cramps

Cramps are common among athletes and usually affect the muscle groups that are being used the most. They settle with electrolyte replacement and rest.

Heat Syncope (Fainting)

This tends to happen when an unacclimatized individual first arrives in a hot climate, usually on sudden standing up or prolonged standing. The normal management for fainting is adequate: the subject lies on his back with the feet elevated.

CHAPTER 3
Fluid Intake

FLUID INTAKE AND EXERCISE

An appropriate fluid-intake strategy may help minimize or delay fatigue. A number of factors may cause fatigue, and a subsequent decrement in exercise performance:

- hyperthermia (high core temperatures);
- dehydration (reduced fluid levels in the body);
- hypoglycaemia (low blood sugar levels);
- glycogen depletion (a reduction in liver and skeletal muscle glycogen stores);
- hyponatraemia (low blood sodium levels);
- unfavourable mental approach (loss of motivational drive).

There are many different types of exercise, which take place in a range of environments – as a result, a particular set of circumstances may be present in one situation and not in another. There are marked differences between individuals in sweat rates, gastric emptying and composition of sweat, and these may result in different fluid requirements. Identification of the possible causes of fatigue in a given situation in heat will guide the exerciser on when it is worthwhile consuming fluid, and what the components of that fluid should be. Consideration must also be given to post-exercise drinks for recovery for the next training session, or to help prepare for the next round in a tennis tournament, or for the final of a running competition. It is suggested that the composition of the fluid should be based on a number of factors. Given the varied factors that may contribute to fatigue plus the different intensities and timescales of exercise, it is unlikely that all of the factors will be present at the same time for a given exerciser, so it is not possible to recommend one solution for all situations. Every individual should make an assessment of his own fluid needs and other factors that may influence performance, and a judgement as to whether fluid intake is necessary and beneficial for a particular event.

Benefits of Fluid Intake

In his review, Coyle (2004) states that dehydration of 1–2 per cent decrease in body weight has no significant effect on an endurance performance of less than 90 minutes in an ambient temperature of 20–21°C; however, a 2 per cent decrease in body weight does have an adverse effect on endurance performance over 90 minutes. Dehydration of 2 per cent body weight in an ambient temperature of about 31–32°C compromises endurance performance in events that last longer than 60 minutes (Coyle, 2004). Thus, the endurance athlete may be able to allow a fluid decrease of 2 per cent body weight in temperate climates without a resultant performance decrement, whereas in a hotter environment, it may be prudent to avoid a 2 per cent decrease in body weight. There is no evidence to suggest that it is essential to replace all fluid loss in order to optimize performance.

An athlete does not adapt to fluid deprivation by being able to 'learn' to cope with less fluid. Athletes can adapt to heat but one of the adaptations involves an increase in sweat rates. As a result, the athlete needs more fluid during exercise when he becomes accustomed to exercising in heat.

In cool conditions it may be more desirable for the endurance athlete to increase the carbohydrate content of the fluid, as carbohydrate depletion will promote fatigue in some endurance events. However, in very hot and humid conditions there will be a great need to replenish fluid loss, and large concentrations of carbohydrate may limit gastric emptying, and delay fluid reaching the cells. This will not help counter the adverse effects of dehydration. Carbohydrate intake will not prevent a rise in core temperature.

It is important that the athlete does not 'over-hydrate', replacing more water than is lost. Disadvantages of this practice include an increased body weight (the athlete has more to carry) and hyponatraemia (*see* below). Noakes (2001) is critical of some of the American College of Sports Medicine (1996) position stand statements on fluid replacement. For example, he questions the need for 'high' volumes of fluid replacement during exercise, and emphasizes that it may also be difficult to replace fluid loss during exercise when sweat rates exceed 600ml/hr. (*See also* Chapter 2.)

Fluid intake has been shown to improve performance in heat, although enhanced performance has not been demonstrated in all studies. There are a number of possible explanations for improved performance with fluid intake. Fluid intake can reduce the adverse effects of dehydration, and enhance endurance performance. One study showed that subjects who took no water could only exercise for 70 minutes, but taking a glucose electrolyte mixture produced an exhaustion time of 91 minutes (Maughan et al., 1989). In this study, endurance time was shortest when no water was consumed, while consumption of a glucose and electrolyte mixture resulted in the longest endurance times. The inclusion of carbohydrates in a fluid replacement beverage has been shown to enhance endurance performance, and may also allow the athlete to sprint faster in the latter stages of exercise.

Fluid Intake During Exercise

Fluid intake during exercise has the potential to decrease heart rate and core temperature. These reductions are likely to enhance endurance performance.

The individual who is dehydrated before exercise are likely to have a compromised thermoregulatory system, reduced performance, lowered skin blood flow and sweat rate. The work of Montain and Coyle (1992) highlights that dehydration has an adverse effect on a number of variables (*see* Figure 7). Eight cyclists exercised in hot conditions (32°C, 50 per cent relative humidity) at between 62 and 67 per cent of maximal oxygen uptake (VO_2max) for 2 hours. On different days they received no fluid, or small, moderate or large volumes of fluid. The authors emphasized that the increases in core temperature and

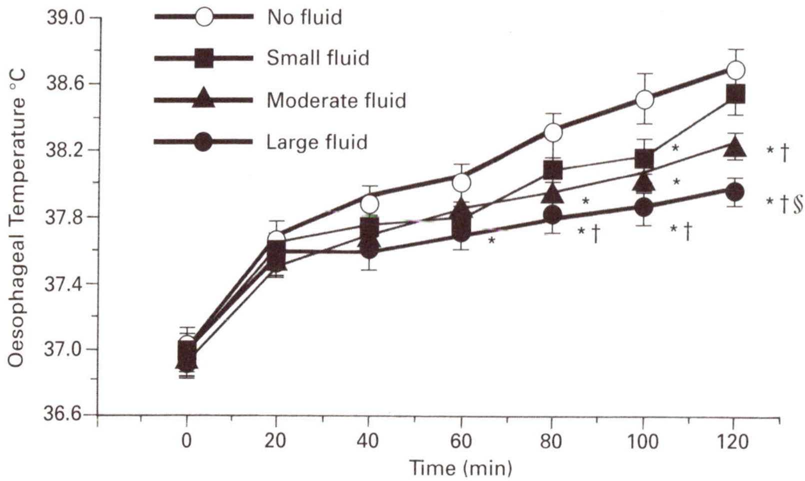

Fig 7. Oesophageal temperature over 120 minutes of cycling in heat.

* Core temperatures were significantly lower than with no fluid
† Core temperatures were significantly lower than with a small volume of fluid
§ Core temperatures were significantly lower than with a moderate volume of fluid intake.

(Adapted from Montain, S. and Coyle, E. (1992), Influence of graded dehydration on hyperthermia and cardiovascular drift during exercise, *Journal of Applied Physiology*, 73, 1340–50. Used with permission from the American Physiological Society)

heart rate, and decrease in stroke volume, were proportional to the degree of dehydration during a two-hour exercise bout. In addition, the level of ratings of perceived exertion (RPE) was highest with no fluid.

Starting exercise in the heat in a dehydrated state is inappropriate, unnecessary and often a result of negligence. A clear strategy needs to be planned to ensure that the athlete is appropriately hydrated before exercise.

Possible Adverse Consequences of Fluid Intake

- Slow athletes in long-distance events are vulnerable to hyponatraemia. Being encouraged to drink large volumes of fluid has resulted in many athletes consuming *too much* fluid. Drinking too much fluid may cause death.
- Some runners stop at drink stations or slow their pace dramatically to allow them to drink comfortably. Poor drinking technique and open cups contribute to a loss of time and limited fluid ingestion.
- Gastrointestinal upset is a possibility particularly with some solutions, such as fructose.
- While cyclists may be able to tolerate high volumes, runners may experience discomfort with high fluid volumes sloshing about in the stomach.
- Large concentrations of carbohydrate (more than 60g/hr) can result in a delay in gastric emptying.

CONSIDERATIONS FOR FLUID INTAKE

Gastric Emptying

The exerciser may drink fluid but, to be of benefit, it must leave the stomach, be absorbed in the small intestine and be taken up by the tissues. Factors that influence gastric emptying include the volume in the stomach, and the concentration of carbohydrate of the fluid. It has been shown that the greater the volume, the greater the gastric emptying up to 600ml of fluid. Thus, within comfort, the exerciser should try to maintain high volumes of fluid in the stomach. Pursuit of this strategy could result in over-drinking and possible hyponatraemia.

Carbohydrate concentrations of more than 8 per cent have been shown to slow gastric emptying, but increasing the glucose concentration of a drink will increase carbohydrate provision. In some situations, it may be desirable to emphasize carbohydrate intake and thus take carbohydrate above 8 per cent. If the emphasis is on water replacement, the carbohydrate content in the drink should be limited to no more than 8 per cent.

Intensities up to 70 per cent of maximal possible VO_2max do not have a major effect on gastric emptying, but intensities above this level reduce gastric emptying.

Sodium

Sodium in the small intestine will enhance fluid and carbohydrate absorption. If sodium has been ingested in recent meals, there will be adequate sodium in the small intestine to promote absorption and there will be no need to include sodium in the drink.

Water follows sodium, so the inclusion of sodium in a drink helps retain water, and has the potential to promote thirst and therefore increase fluid intake. The ingestion of plain water lowers the osmolarity of the extracellular fluid (ECF), which results in a decreased impetus to drink and an increase in urine output. It is for these reasons that a rehydration drink should contain sodium. In events where there is a high level of sodium loss (in excess of 3–4g), or in exercise bouts of longer than 2 hours, it is recommended that sodium should be included in the drink (Coyle, 2004).

Individual Variability

There are many different types of exercise in a range of environments as well as different intensities and timescales of exercise. Between individuals there are also marked differences in sweat rates, speed of gastric emptying, rate of intestinal absorption, sweat rates and composition of sweat. Therefore, general fluid intake guidelines have limitations for the individual athlete. Each athlete should determine his own fluid loss in training in a range of environmental conditions to identify the rate of fluid intake that is appropriate to him. An individual's fluid loss and the volume of fluid that needs to be replaced can be estimated as follows (giving an example for a 2-hour exercise session):

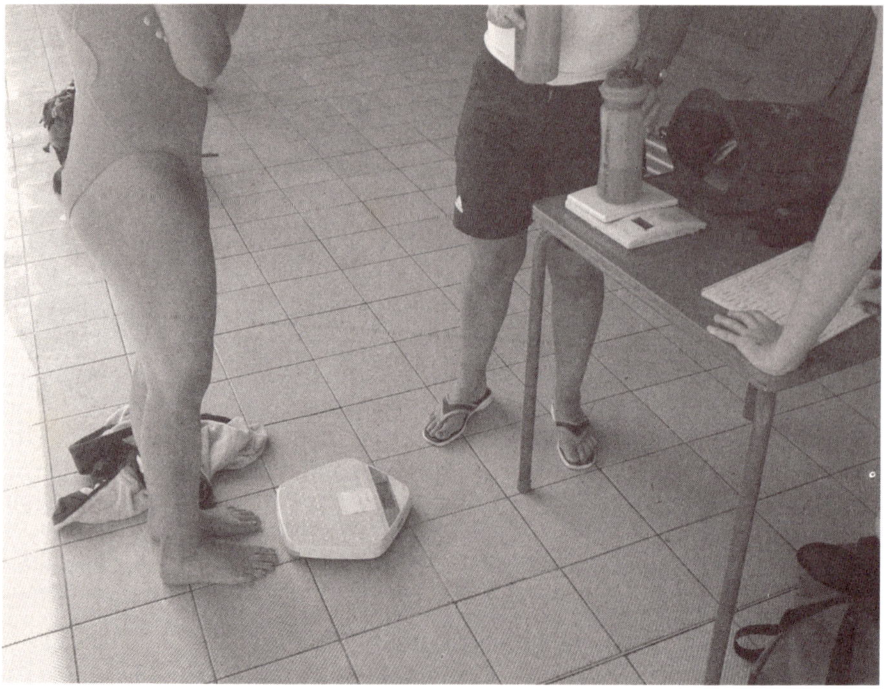

Fig 8. Athlete being weighed in standardized clothing on the same scales to determine fluid loss and the volume needed for adequate fluid replacement. Fluid intake is also included in the calculation.

(Reproduced with permission from Andrew Murray, The National Stadium, Sports Health & Injury Clinic, Hampden Park, Glasgow)

1 Measure body weight (dry, wearing only underwear) before exercise begins (70kg).
2 Measure fluid intake during the exercise bout (1ltr).
3 Measure body weight (dry, wearing only underwear) at the end of exercise (68kg).
4 Calculate total fluid loss by equating each kg of body weight lost to 1ltr of fluid (2kg lost + 1ltr intake = 3ltr).
5 Calculate sweat rate by dividing total fluid lost by length of exercise (3ltr/2 hours = 1.5ltr/hr).

PRE-EXERCISE FLUID

Prior to the start of exercise, the athlete should be 'topped up' with fluid and ensure that his urine is lightly coloured. In most situations the exerciser has no excuse for starting exercise in a dehydrated state. Athletes who have just recently arrived in a hot climate, and who are involved in multiple exercise sessions such as heats or training sessions, are particularly vulnerable to the dangers of starting an exercise session dehydrated. Recommendations from the American College of Sports Medicine (ACSM, 1996) are that 400–600ml of water should be ingested 2 hours before exercise to ensure that the athlete is fully hydrated. As soon as fairly high-intensity exercise is started, urine production will decrease and possible fears of the need for a toilet stop will disappear. Some researchers suggest that fluid should be consumed in the few minutes before the event. This procedure should be practised to ensure the athlete has no stomach discomfort.

FLUID DURING EXERCISE

Identifying Particular Needs

It is worth re-emphasizing here the need for self-assessment of fluid needs, since sweat rates, electrolyte content of sweat, taste and gastric emptying can vary widely between individuals. Athletes should interpret general fluid guidelines with caution. Dugas et al. (2005) consider that athletes should consume fluids *ad libitum*, meaning according to thirst. They refer to guidelines from a range of organizations, ranging from 400–800ml/hr to 1,800ml/hr for fluid intake during exercise, but they consider that the *ad libitum* approach is sensible as it mirrors that of creatures who have not been educated to adopt other strategies. The American College of Sports Medicine (2005) advises that athletes should drink early and regularly. They caution that reliance on the perception of thirst will result in incomplete restoration of fluid losses resulting from sweating. Athletes should be familiar with their personal typical sweat rates in a given environmental condition (*see* above) and take the necessary fluid on board to replace lost fluid. 'Heavy sweaters' who are large and run fast in very hot conditions, for example, are likely to need particularly large volumes of fluid.

It is important for the athlete to avoid both dehydration and hyponatraemia (*see* below), and either too little or too much fluid will be inappropriate. Prior experience of exercise in heat, and experimentation in training and in events that do not matter, will help the athlete to determine the appropriate fluid intake for him. Possible gastric upset can be identified in training before competition. Athletes should also practise drinking, experimenting with the taste of various fluids, and whether he can drink without major spillage while on the move.

Short-Duration, High-Intensity Exercise

During high-intensity short-duration exercise, such as a 1,500m run, fluid intake is not worthwhile; dehydration should not be a problem for the athlete who starts the event fully hydrated. In addition, fluid that remains in the gastrointestinal tract will not have a beneficial effect. The fluid is inside the body but it is in a tube and could be deemed to be outside the body as it has not reached the cells.

In events that last for less than 40 minutes, beneficial effects of fluid intake as measured by decreases in heart rate and core temperature are not likely to manifest themselves. It is likely that exercise will have to last from 40 minutes upwards before a decreased heart rate and core temperature will be evident as a result of fluid intake. The extra time is needed to allow the fluid to reach the cells.

It has been recommended by some that the exerciser should try to consume 500–1,000ml water to replace lost sweat and minimize the increase in core temperature during high-intensity exercise of around an hour (Gisolfi and Dutchman, 1992). However, not only are top level athletes unlikely to break their rhythm by picking up fluid, but there may be problems with spillage during high-speed running, and limited gastric emptying with high-intensity exercise resulting in possible discomfort and also relatively little water will be absorbed.

Medium to Long-Duration Exercise

It is suggested by Coyle (2004) that, in exercise lasting from 1 to 3 hours, fluid intake should match sweat rate where possible, but that fluid intake should not *exceed* sweat rate. The American College of Sports Medicine (ACSM) (1996) recommends that the exerciser should drink often and regularly during exercise, and should not rely on thirst as a guide for fluid replacement. It has been shown that individuals, who use thirst to stimulate their drinking, normally only replace one-half to two-thirds of their fluid loss. When the use of the thirst mechanism fails to replace fluid loss, it is called 'voluntary dehydration'.

The ACSM (1996) recommends a range of fluid volumes that the exerciser should consider – between 150 and 350ml of water every 15–20 minutes. Some have questioned the advisability of prescribing specific volumes, counselling that each individual should determine their own appropriate volume and concentrations, as a result of trial and error during training sessions. For example, an athlete could start by trying 600ml/hr in training, and then adjust intake accordingly. Noakes (2001) recommends that fluid intake rates are driven by the athlete, who can decide how much fluid should be taken. He states that his recommendation normally results in a fluid intake of between 400ml and 800ml per hour, and that this range applies to a range of activities and individuals. Observation of running races shows that some elite athletes take in very small fluid volumes. This may be a result of the fact that fluid leaves their stomach at a slow rate. The pounding of the running promotes an uncomfortable feeling in the stomach. In contrast, elite cyclists may be able to consume much higher volumes despite the fact that they are exercising at high intensities.

The fluid should contain a small amount of sodium and chloride and no greater than 8 per cent carbohydrate. Coyle (2004) recommends that sodium of around 20–40mmol/ltr should be ingested during exercise particularly when events last more than 1 hour. However, the necessity and effectiveness of adding sodium to combat hyponatraemia is contested by Hew-Butler et al. (2006) (*see* page 59). Carbohydrate consumption of 30–60g/hr is recommended when the exerciser is involved in high-intensity continuous or intermittent activities that last more than 1 hour. It is strongly advised that the carbohydrate solution should be taken regularly with the aim of maintaining a continuous even flow of carbohydrate to the bloodstream. Noakes (2001) suggests that carbohydrate concentration should be increased in the latter stages when the replacement of carbohydrate is very important and when the athlete is less likely to want to drink.

The literature indicates that it may be possible to meet the demand for fluid and carbohydrates. It is possible to ingest between 600 and 1,000ml/hr of fluid and still include the recommended carbohydrate concentration. For example, a 750ml volume of fluid with a concentration of 8 per cent carbohydrate (including sodium of 20mmol/ltr) consumed in 1 hour will provide 60g of carbohydrate per hour. The rationale for the above includes the need to replace fluid, and to facilitate water absorption by using a sodium/chloride and carbohydrate mixture. The addition of carbohydrate will help counter the adverse effects of carbohydrate depletion. The American College of Sports Medicine (1996) stresses that carbohydrate has advantages as it will maintain blood-glucose levels and promote the use of carbohydrate as a fuel for exercise.

Long-Duration Exercise (over three hours)

In exercise of long duration (over 3 hours), it is important to ensure that the volume of water ingested does not exceed that of fluid lost through sweating. Athletes must not increase their body weight by taking on excess fluid during exercise. This will not only be disadvantage in weight-bearing activities such as running, but may also result in the possibility of hyponatraemia (*see* below). The exerciser who has a water intake of 750–1,000ml per hour for a duration of over 5 hours is in danger of developing hyponatraemia. The guidelines on carbohydrate provision for medium-duration events also apply to events of a longer duration.

RECOVERY FROM EXERCISE

Many individuals carry out more than one exercise bout on the same day. They may be involved in heats, they may play more than one squash game per day, or they may train at least twice per day. They may also train 14 times per week. It is important to ensure that such individuals are properly hydrated for their next exercise challenge, and that the appropriate fluid is used.

Fluid and electrolyte loss and carbohydrate depletion are common occurrences after exercise. Some have suggested that rehydration drinks should have a similar sodium concentration to the fluid lost in sweat. As the sodium content of sweat may range from 20 to 80mmol/ltr, it is difficult to provide general guidelines. Nevertheless, it is clear that plain water as a rehydration drink has major limitations. The ingestion of plain water reduces the plasma omolarity, decreases the urge to drink, and results in an increased urine production; all combine to slow rehydration.

The main consideration for a rehydration drink is replacement of water and electrolyte losses from sweating and carbohydrates. It is essential to include sodium in the rehydration drink, as this will allow rehydration to take place faster. Research by Maughan and Leiper (1995) showed that fluids with higher sodium concentrations had lower urine outputs from 1.5 hours post-exercise and beyond (Figure 9). They dehydrated six males on four occasions, then used

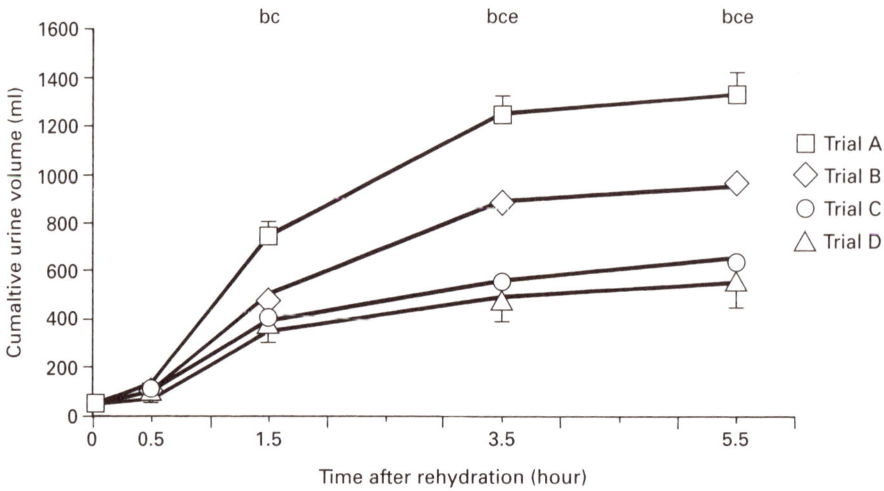

Fig 9. The group that had the drink with the highest sodium concentration had the lowest urine output and the highest fluid retention.

(Maughan R. and Leiper, J. (1995), Sodium, intake and post-exercise rehydration in man, *European Journal of Applied Physiology*, 71, 311–19.) The sodium contents of the drinks were as follows: A: 2mmol/litre; B: 26mmol/litre; C: 52mmol/litre; D: 100mmol/litre. (Permission given by Springer-Verlag Gmbh, Tiergartenstr.17, 69121 Heidelberg, and Professor Ron Maughan, University of Loughborough)

drinks of varying sodium concentrations for rehydration. The higher sodium concentrations enhanced fluid retention.

The inclusion of sodium will also replace sodium lost through sweating. Maughan (1991) recommends that the rehydration drink should have a sodium content of 40–80mmol/ltr. Many sports drinks have a sodium content of around 10–25mmol/ltr, and some soft drinks such as lemonade may have a sodium content of only 1–2mmol/ltr, whereas fluids that are used to treat diarrhoea have a sodium content of around 80mmol/ltr. One possible disadvantage of a sodium content of 80mmol/ltr is the possible inhibition to drink such salty fluids.

The fluid that is consumed should be larger than that lost through sweating to allow for the ongoing essential fluid loss via urine loss.

Even though the next exercise bout may be only a short time away (for example, several hours), the inclusion of carbohydrate is likely to promote performance enhancement. This may be accomplished most easily in fluid form but some may find it easy to consume bread and potatoes fairly soon after exercise. The rate of synthesis is not influenced by the choice of liquid or solid. It is stressed that muscle glycogen re-synthesis is faster in the first hour after exercise than at later times; athletes should take carbohydrate as soon as possible, to take advantage of this, and to provide as much time as possible for the process to take place. If no carbohydrate is consumed in the first few hours after exercise, the athlete will have very limited restoration of glycogen, which may limit

Glycogen re-synthesis

- Muscle glycogen re-synthesis is faster in the first hour after exercise than at later times.
- Failure to take carbohydrate in the recovery phase results in very limited restoration of glycogen, which may limit performance in training and competition.

performance in training and competition. This advice is particularly true for an exercise bout that will take place within 8 hours. When the next exercise bout is not for 24 hours or more, the athlete can organize carbohydrate intake into a structured meal programme.

During the recovery phase after exercise Burke et al. (2004) recommend that athletes should consume sufficient carbohydrates so that they have adequate fuel for their next training session. The paper by Burke et al. (2004) provides much more detailed information on this topic.

It is important that athletes and coaches are aware of the need for rehydration, and the replenishment of carbohydrate and sodium. Steps should be taken to make fluid and foodstuffs readily available as soon as the athlete stops exercising. For example, the provision of fluid and food in a changing room or a socializing room has the potential to enhance compliance with recommendations. The social area should be cool and comfortable, and could include posters that highlight the need for the appropriate replacement fluids. A choice of drinks should be available and these should be considered to be palatable by the athletes.

ENHANCING FLUID INTAKE

During Training

Practice of fluid intake during training has a number of advantages. The exerciser will be able to determine the appropriate temperature and flavouring of the drink to enhance palatability, and encourage a higher fluid intake. Monitoring of body weight will provide information on fluid loss, help identify if the subject is a 'heavy sweater', and give an indication as to appropriate fluid intake volumes. Practice will also help the exerciser devise strategies to increase the volume that is actually ingested and not spilled. Athletes should not drink from an open cup; a closed bottle will prevent spillage. It is also possible to use a drinking straw held between the fingers of the hand that is over the top of the cup.

If long training runs or cycle rides are planned, it is sensible to make sure that 'waterholes' can be found en route. It may be possible to use a circuit where the athlete can return home for fluid and/or plan the route so that water fountains are available, or bottles are hidden in strategic locations. Some fluid may be carried, or helpers could provide fluid.

Fig 10. Marathon runners drinking from a cup.
(Reproduced with permission from Trevor Graham, University of Glasgow Photographic Unit)

Developing a Routine

A clearly defined routine for fluid intake should be implemented. The inclusion of a water bottle in the athlete's kit bag while the athlete travels, and during the stay in the hot environment, will help the athlete meet fluid needs. An insulated bottle that keeps the fluid cool will promote fluid intake. The need to consume large volumes of fluid may result in financial and time pressures for the athlete. The athlete should devise ways of providing a regular uninterrupted routine to access fluid without associated difficulties, which will inhibit fluid intake.

Practising drinking in training will allow the individual to develop techniques to avoid major spills, and promote an awareness of the problems of drinking on the move. Experimentation will help the athlete to determine the desirable characteristics of the drink that particularly suits him. The inclusion of sodium and flavouring may enhance palatability and encourage drinking.

Studies indicate that palatability influences how much athletes will drink, and that flavoured water and orange juice enhance fluid intake compared with plain water.

While evidence suggests that gastric emptying is not greatly influenced by differences in the fluid temperature, for the athlete who is exercising in hot conditions, drinks are likely to be more palatable when they are cool. The fluid temperature should be less than the ambient temperature. The American College of Sports Medicine (1996) recommends fluid temperatures between 15° and 22°C.

Fluid should be easily available. It should be placed in cool boxes next to the track or pitch when necessary and should be easily accessed during meals. Personalizing bottles reduces the likelihood of cross-contamination and allows quantification of the volumes drunk by each individual.

Remember to keep the drinking end of the bottle clean!

FLUID INTAKE FOR TEAM GAMES PLAYERS

Studies on soccer and rugby players have shown that fluid loss and sweat composition vary greatly between players. It has been suggested by some that fluid

Fig 11. Drinking from a water bottle prevents spillage.

(Reproduced with permission from Andrew Murray, The National Stadium, Sports Health & Injury Clinic, Hampden Park, Glasgow)

Fluid intake strategies

- Determine the appropriate drink for each situation by evaluating frequency, duration and intensity of exercise, and recovery.
- Determine the factors that may contribute to fatigue.
- Avoid dehydration before you reach your destination by drinking fluid regularly during travel and by avoiding alcohol, which promotes urine production.
- Carry a water bottle while travelling to the hot location, and during your stay there, and have a water bottle by your bedside.
- Devise a plan to enhance fluid intake.
- Establish a clearly defined routine for all situations for fluid intake in hot environments, including a strategy for drinks during meals, training and competition, making fluid easily accessible.
- Avoid starting exercise in a dehydrated state. In most cases it is inexcusable to start exercising dehydrated.
- Take between 300–500ml of fluid 2 hours before exercise.
- Fluid should be palatable and at the correct temperature.
- Flavouring and a cool fluid temperature will improve palatability and, therefore, enhance fluid intake. Use an insulated water bottle and provide cool boxes at the training site for water bottles.
- During training and competition, fluid should be available during the warm-up, breaks in games and training, in the dressing room and in the post-exercise social areas.
- Establish regular drink breaks during training sessions.
- Measure body weight before and after exercise sessions and fluid intake during exercise to determine fluid losses and give guidance for fluid replenishment.
- Do not increase body weight during exercise by drinking more than the volume of fluid lost through sweat loss.

losses in a cool environment are similar to fairly warm environments. Possible explanations for these suggestions include the fact that intensity may be reduced in hot conditions and more clothing may be worn in cool conditions. Average sweat loss for soccer players can be expected to increase from 0.85ltr per game in a cool environment to over 2ltr in hot conditions, while some individuals may experience around 3ltr of sweat loss. It appears that soccer players only replace about a third of their sweat losses in hot conditions with only a third to half of sweat loss being replaced at half time. Decreases in body weight of 2 per cent and above due to dehydration may result in decrements in mental functioning that could have adverse effects on the games player who has to make many decisions quickly. Bearing in mind the above, a re-assessment of hydration strategies in soccer players may be warranted.

Given the individual variability in sweat loss and wide-ranging environmental conditions, it is difficult to provide clear-cut guidelines for fluid intake for games players. Players are likely to tolerate greater levels of dehydration in cool conditions but they will be more susceptible to training and match fluid loss if they begin a session dehydrated. Some suggest that a fluid loss of 2 per cent body weight may be acceptable in cool conditions. The inclusion of carbohydrates has the potential to enhance skill performance in sports such as soccer and tennis. Carbohydrate depletion is a possible cause of fatigue in soccer and other team games such as rugby and hockey (*see* Chapter 2).

Assessment of and recommendations for individuals are necessary. For example, 'heavy sweaters' should be identified. Also second row forwards in rugby are normally very large, and will probably need larger volumes of fluid, compared with a more diminutive scrum half. An evaluation of the training, match and tournament climatic conditions should determine the appropriate guidelines for each player.

Players should avoid starting training or competition dehydrated by consuming 300–500ml of water in the 2-hour period before activity begins. Regular fluid intake should be made easy for the players by identifying when they will be able to drink without disruption or inconvenience. Fluid should be readily available in the dressing room, on the substitutes' bench and on the field of play during stoppages such as injuries and corners. Bottles should be labelled for each player and, if possible, helpers should be organized so that the players are brought fluid on a regular basis.

Appropriate fluid intake should be promoted, with hydration posters clearly displayed in the dressing room, and coaches reminding players about regular drinking, and providing regular and official drink breaks. Players should also be encouraged to monitor their hydration status and weighing should be part of the normal routine. It should be made easy for the players to weigh themselves, whether at home or away.

Where possible, in hot conditions, the players should drink during the warm-up, between the warm-up and the game, and during the game when the situation allows. They should include carbohydrate and sodium in the drink. The fluid should be kept in cool boxes and be flavoured to promote uptake. Post exercise, the players should be exposed to an environment that allows for easy rehydration – appropriate fluid should be available in the dressing room and at post-activity social areas.

MONITORING HYDRATION STATUS

Light-coloured urine (crystal-clear or very pale yellow) is a sign that hydration status is good, whereas a dark, straw-coloured urine indicates a dehydrated state. Certainly, if the urine is rich gold before the event, the medal is unlikely to be gold. (On the other hand, continuously crystal-clear urine is not a target to aim for since it may indicate over-consumption of water.) 'Pee charts' allow the exerciser to compare urine colour with a colour scale – the darker the colour, the greater the degree of dehydration – and indicate whether he needs to take on more fluid. It is also possible to measure urine volumes but this procedure can be irksome, and compliance may be low. Another possible strategy is to be aware of the normal times when an athlete uses the toilet – first thing in the morning, mid-morning, at lunch and so on – and suspect possible sub-hydration if one of the usual visits is missed.

The monitoring of body weight provides a reasonably sensitive marker of hydration status, with 1ltr of water equivalent to approximately 1kg. Some guidance is needed on how to carry out the weighing, and on how to interpret the results. The athlete should use the same set of scales, which should be placed on a flat surface and, when possible, in a room of similar temperature regardless of location. Body weight should be measured in standardized conditions. For example, body weight should be taken each morning after the athlete has risen out of bed and visited the toilet. The athlete should wear the same clothing (usually just underwear) on each occasion. If the athlete wishes to monitor fluid loss as a result of a game or training session, measurement of body weight should again be standardized and the athlete should wear underwear and be dried off before taking body weight after the exercise bout. If the athlete is aware of his normal weight, a meaningful evaluation of hydration status can be made. Any training diary should include information on the athlete's body weight.

In any sportsperson, rapid change in body weight, either up or down, is due to changes in body water. However, it should be noted that slower changes in body weight might reflect factors other than differences in fluid balance. A decreased appetite, as a result of living in the heat or dislike of a new diet, may be the cause of weight loss. Conversely, an increased energy intake resulting from a large increase in fluid that contains carbohydrate may result in an increase in body weight.

Hydration check

- Monitor volume and colour of urine. Try to have large volumes of clear/pale yellow urine, but be aware that continuously clear urine may be a sign of over-hydration.
- Monitor body weight in standardized conditions – 1kg loss is equivalent to approximately 1ltr of fluid.
- Take a pair of scales with you so that you can monitor body weight changes in training to determine fluid loss and have information on the appropriate fluid intake.

Table 3 **Monitoring hydration status**

Hydration status check	*OK*	*Dehydrated*
Observe urine	Crystal-clear to pale yellow	Dark yellow to brown
Toilet frequency	Similar frequency to home	Reduced frequency
Weight	Steady body weight	Loss of body weight

HYPONATRAEMIA

Exercise related hyponatraemia with encephalopathy was first reported by Noakes in 1981 and then Frizzell et al. (1986) and has normally been associated with ultra-marathons. The incidence of hyponatraemia in long-duration events of over 6 hours is reported to be as high as 27 per cent, compared with 5–9 per cent of marathon runners (Hsieh, 2004). Based on their sample of 488 runners in the 2002 Boston marathon, Almond et al. (2005) estimated that of the almost 15,000 finishers, around 1,900 could be classified as being hyponatraemic (13 per cent had a serum sodium concentration under 135mmol/l), with about 90 finishers (0.6 per cent who had a serum sodium concentration less than 120mmol/l) being categorized as having critical hyponatraemia.

A number of risk factors for hyponatraemia have been identified. In the Almond et al. (2005) study, weight gain during the marathon, intake of more than 3ltr of fluid during the marathon, a race time >4 hours, female sex and a low body mass index were the main variables associated with hyponatraemia. The authors concluded that weight gain during the marathon was the best predictor of hyponatraemia. The increase in body weight during the race was attributed to an excessive fluid intake. Slow athletes may be susceptible to hyponatraemia. For example, slow runners produce less heat than fast runners, and therefore may not need a large fluid intake, but have a long period of time in which to consume fluid. Females tend to be slower and may have less need for fluid than males as they are generally much smaller. The athlete who is a 'heavy sweater', losing a great deal of salt, and has a restricted salt intake in his diet may be particularly susceptible to hyponatraemia.

A recent mathematical study (Montain et al., 2006), confirms the early clinical studies by Noakes (1981) and Frizzell et al. (1986), that the major cause of hyponatraemia is over-consumption of plain water but that there may be other factors. Interesting comments have been made on the validity of these other factors (Noakes, 2006; Noakes and Speedy, 2006; Weschler, 2006; Murray, 2006; Armstrong, 2006).

Participants in long-duration exercise should weigh themselves before and after lengthy training sessions to help them titrate their hydration and fluid strategy more precisely. The phenomenon of hyponatraemia may be a side-effect of the over-emphasis on the dangers of dehydration during long-duration exercise, with people believing that they must consume water at every opportunity. Athletes must avoid weight gain during exercise as a result of fluid intake. In particular, high-risk athletes must not take large volumes of fluid. For example, 'slow' female runners (about 10km/h) who drank 1ltr per hour for

Avoiding hyponatraemia

- Identify those most at risk – those who consume large fluid volumes resulting in fluid overload, slow athletes, females.
- Do not gain weight via fluid intake during exercise.
- Do not over-emphasize fluid intake.

4 hours were described by Twernbold et al. (2003) as 'likely to develop fluid overload'. Dugas et al. (2005) highlight cases in which athletes developed hyponatraemia with fluid intakes of around 750ml/hr over 12 to 13 hours in Ironman triathlons. They stress that these volumes are not excessive and are within some guidelines for fluid intake.

The ingestion of large volumes of plain water during exercise should be avoided. The study of Hew-Butler et al. (2006) demonstrated that the inclusion of ad libitum sodium during an Ironman triathlon was not needed to maintain serum sodium concentrations and had no effect on triathlon performance. They divided competitors in an Ironman triathlon into three groups. One took salt supplements, the second took placebo and the third took no supplements. At the end of the race there was no difference in serum sodium levels in the three groups. They suggested that a daily intake of 1.5g sodium in the diet is adequate and there is no need for the athlete to consume more.

Failure to recognise hyponatraemia and/or incorrect treatment with hypotonic fluid can be fatal. However, rapid recovery occurs if hypertonic (3–5 per cent) saline solution is infused (Noakes and Speedy, 2006). Noakes highlights that the over-consumption of some sports drinks may result in a reduced plasma sodium and hyponatraemia.

WATER ON THE SKIN DURING EXERCISE

Spraying water on the exercising athlete has been shown to have a limited cooling effect and may even promote problems. Sodden vests can cause bleeding nipples, and wet shoes and socks may result in blisters and unnecessary discomfort. However, the application of cold water to the skin, either by

What and how to drink?

- Determine which drink is suitable and which volumes are appropriate by practising and experimenting with fluid intake during training. Do not experiment in a competition.
- Ensure palatability – if the fluid tastes 'good', the exerciser will find it easier to take it in.
- Find a system that prevents spillage – use bottles with nozzles or drinking straws through fingers.
- Fluid should be easily accessed at all times (at meal times, in the training area) to facilitate fluid intake.
- Make sure that enough fluid is consumed to meet the demands of living and training in a hot environment.
- Ensure that fluid is safe. Avoid tap water and ice in 'high-risk' countries. If necessary, purchase commercial bottled water.
- Take care that the energy intake from increased fluid consumption does not promote weight gain.
- Heat-acclimatized individuals have enhanced sweat rates, and thus need more fluid than the non-acclimatized.

spraying or sponging, can result in constriction of blood vessels (if skin cooling takes place) and a reduction in the blood pooling. Applying cold water to the arms and legs, in order to cool the skin, will reduce blood pooling in these areas. The enhanced venous return will allow a higher stroke volume and a reduced heart rate to be maintained.

Rehydration

- Rehydration for the next exercise bout should start at the end of the previous exercise session.
- Return to pre-exercise weight by next training session.
- Rehydration drinks after competition and training should address the need for replenishment of water, electrolytes and carbohydrates.
- The rehydration drink should include sodium to help reduce urine output and enhance fluid retention.
- Glycogen re-synthesis is most effective in the first hour after exercise.

Don'ts

- Throwing water over the body has minimal cooling effect. Drink the fluid instead. However, during prolonged exercise in the heat, pouring cold water over the head can provide a great feeling of comfort.
- Don't attempt to adapt to water deprivation or dehydration – you may make yourself ill!
- Don't consume alcohol and large amounts of tea and coffee, as this will enhance urine production.
- Don't over-hydrate – avoid weight gain via fluid intake during exercise.

CHAPTER 4
Heat Acclimatization

Young (1986) makes the following distinction between acclimatization and acclimation:

- *acclimatization* is an adaptation as a result of changes to the natural environment, as in summer weather, or when living temporarily in a hot climate;
- *acclimation* is an adaptation induced by unusual alterations in environmental conditions, for example, exposure to a sauna, botanical gardens or an environmental chamber.

According to Sawka and Pandolf (2001), heat acclimatization 'results in biological adaptations that reduce the negative effects of heat stress'.

Endurance performance is likely to be reduced in heat. Athletes who have not recently exercised in hot conditions may have an increased perception of fatigue, and experience lethargy compared with exercising in cool environments. Heat acclimatization can reduce to a certain extent the detrimental effects of heat. Heat acclimatization improves endurance performance, reduces thermal and circulatory strain, enhances cognitive performance (making the athlete better able to think), and lowers the subjective stress resulting from heat.

Heat acclimatization improves endurance performance in heat. Athletes from temperate climates who are exposed to heat and/or train in heat can expect to improve in relatively short periods of time, with major improvements in the first week of exposure. For example, athletes will be able to do much more exercise in heat and may be able to complete exercise bouts that would not be possible without heat acclimatization. In one study twenty-four males were asked to walk for 100 minutes on successive days in dry heat. On day 1 no subject was able to complete the 100-minute walk in an ambient temperature of 49°C and 20% humidity but by day 3 of the acclimation period six subjects were able to walk for 100 minutes. After 7 days all but one subject could successfully complete the walk. It should be noted, however, that heat acclimatization may not allow the individual to perform as well in heat as in cool conditions.

PHYSIOLOGICAL CHANGES WITH HEAT ACCLIMATIZATION

Reduced Physiological Strain

After heat acclimatization, a number of adaptations result in a reduced physiological strain in heat. There is an increase in sweating, a reduced thermal strain, greater stability in the cardiovascular system and a more favourable fluid balance.

Sweating

Following heat acclimatization, a number of changes to sweating enhance the cooling capability of the athlete and facilitate heat loss, resulting in reduced thermal strain. Not only does sweating start sooner after heat acclimatization, sweating capacity is also higher and at a greater rate. The ability to sustain sweating at higher rates during prolonged exercise sessions in heat is increased. With acclimatization in humid heat, there is more sweating on the limbs resulting in an increased surface area for sweating and thus an increased potential for more sweat to evaporate and facilitate cooling. There is an enhanced sweating sensitivity – in other words, an increased response to temperature changes. The salt loss in sweat is reduced and, as there are fewer impurities in the sweat, the evaporation of sweat is increased. The reduced sodium in sweat and thus the greater retention of sodium in the body promotes the sensation of thirst and an increase in total body water.

After heat acclimatization, the athlete sweats more, and this increases his need for fluid. Fortunately one of the effects of heat acclimatization is to promote the desire to drink i.e. the heat acclimatized individual has an enhanced thirst mechanism.

Cardiovascular

Exercise heart rate has been shown to be reduced by around 30 beats/min after heat acclimatization. Most of the heart rate decrease takes place within 4 to 5 days. Possible reasons for a lower heart rate include a reduction in core and skin temperature and an increase in the temperature difference between core and skin temperatures following heat acclimatization. It is hypothesised that the greater core to skin temperature gradient and reduced skin temperature result in less venous pooling and an enhanced venous return resulting in a reduced heart rate. Another possible factor which may contribute to a lower heart rate is an increased plasma volume.

Enhanced venous tone reduces blood pooling. This, plus an increase in plasma volume and total body water, helps maintain blood pressure, and this in turn contributes to a reduced likelihood of fainting in heat. Skin blood flow starts earlier and is increased, and provides greater potential for heat loss. Skin temperature is maintained at a lower temperature.

Metabolic Rate

Metabolic rate is reduced, resulting in a lower heat production and less heat stress. The lower metabolic rate also results in a decrease in energy for a given task.

Core Temperature

Enhanced performance after heat acclimation has been shown to be partly due to a lower core temperature before exercise. Possible contributory factors to a reduced core temperature include a lower resting core temperature and a decrease in the metabolic rate during exercise, resulting in less metabolic heat production. Heat acclimation increases time to exhaustion partly via a lower pre-exercise core temperature. Individuals are less likely to experience heat distress after heat acclimation.

There may be a decrease in muscle glycogen use after heat acclimatization, which will also enhance endurance performance.

The work of Lind and Bass (1963) highlights that heat acclimatization has beneficial effects and that improvements in some physiological variables can be expected fairly quickly. After 9 days of walking in heat for 100 minutes each day, their subjects had falls in core temperature from 39°C on Day 1 to 38°C on Day 9. Heart rate decreased from around 170 beats/min to 130 beats/min and sweat rate increased by around 10 per cent. The authors stated that the above variables had reached a plateau after 8 days of heat acclimatization. Thus with appropriate heat acclimatization, major changes in heart rate and core temperature and sweat rate can be expected in around a week.

The benefits of heat acclimatization are clear:

- Endurance performance is improved.
- Sweating starts sooner.
- Sweating capacity is higher and is at a greater rate. With acclimatization in humid conditions, there is more sweating on the limbs resulting in an increased surface area for sweating and potentially greater volumes of sweat that can be evaporated.
- There is an enhanced sweating sensitivity – sweating starts at a lower core temperature.
- The salt loss in sweat is reduced and consequently the evaporation of sweat is increased.
- Exercise heart rate is decreased.
- Blood pressure is maintained better reducing the likelihood of fainting in heat.
- The enhanced venous tone reduces blood pooling.
- Skin blood flow starts earlier and is increased, and provides greater potential for heat loss.
- The lower core and skin temperatures lessen the need for the redistribution of blood to assist with cooling.
- Confidence to exercise and compete in the heat is improved.
- Individuals experience less heat distress and reduced discomfort.
- There is enhanced cognitive performance (the athlete is able to think better).
- Metabolic rate is reduced, resulting in a lower heat production, and less heat stress. The lower metabolic rate also results in a decrease in energy for a given task.
- There may be a decrease in muscle glycogen use, which will enhance endurance performance.

Aerobic Training

The aerobically trained athlete has some heat acclimatization as aerobic training produces increases in sweating and core temperature. The aerobically trained individual starts sweating sooner than when untrained and also sweats at a greater rate. The benefits of aerobic training are supplemented by heat acclimatization.

The heat acclimatized and aerobically trained individual starts sweating earlier than when unacclimatized but trained. There are also differences in sweat composition depending on aerobic training status and degree of heat acclimatization. The sweat of the aerobically trained person contains less sodium and chloride than the sweat when unfit. The trained and heat acclimatized person has even lower sodium and chloride sweat values than when unacclimatized. The trained individual has some heat acclimatization benefits, but they are not as great as when trained and heat acclimatized.

Examples cited by Noakes (2001) show that the sodium sweat concentration of the trained and acclimatized was 40mmol/L (1.8g/L), trained and unacclimatized 60mmol/L (2.6g/L) and untrained and unacclimatized 80mmol/L (3.5g/L). These sodium values also highlight that sodium sweat concentrations are much lower than the typical plasma sodium concentration of 140mmol/L (6.1g/L) regardless of the status of the subject.

Effect of aerobic training and heat acclimatization on sweat

- The trained and heat acclimatized subject starts to sweat sooner than when only trained. The trained subject starts to sweat sooner than when untrained.
- The sweat concentrations of sodium and chloride are lower in the heat acclimatized and trained subject compared with the trained only state.
- The sweat concentrations of sodium and chloride are lower in the trained subject compared with the untrained state.

ACHIEVING HEAT ACCLIMATIZATION

Methods

The athlete should identify the anticipated environmental conditions and acclimatize by mimicking those conditions. Approximate temperatures and humidity levels at the time of the event should be ascertained so that they can be simulated at home. (For example, will the heat be dry or wet?) Heat acclimatization results from repeated exposure to heat stress where the core temperature and sweating are increased on a regular basis. This can be achieved by exercise, or by sitting in an artificially hot environment such as a sauna. (It is often difficult to spend a long time in a sauna, as it is too hot; reduction of the temperature to tolerable levels will reduce the heat stress and sweating. In addition, dehydration may result in the sauna.) Active acclimatization is more desirable than resting acclimatization since it requires less time and is therefore more economic. The mode of activity is not a critical aspect, as long as the aim of increasing core temperature and sweat rates is achieved.

It is unlikely that the exerciser will be able to perform his 'normal' training in heat initially, and the reduced intensity and duration in the acclimatization sessions could lead to de-training. Quality training sessions should be carried out in cool conditions – early in the morning, in the shade or in an air-conditioned facility – while acclimatization sessions are done in the heat.

The intensity and duration of exercise should be increased gradually. Attempts to accelerate the heat-acclimatization process may expose the athlete to unnecessary danger and lead to heat injury, illness or even death. Some American footballers have died of heat injury when they reported back for training after a close-season break. They exercised in their protective clothing (limiting evaporative heat loss) in high humidity, and had limited fluid intake.

During heat acclimation/acclimatization sessions the athletes should have a plentiful supply of fluid and should be instructed to drink regularly. The use of heart rate recorders should be considered. Regular monitoring of heart rate will not only allow the athletes to regulate exercise intensity easily but may provide guidance on heart rate changes as the heat acclimation/acclimatization process takes place. A reduction in heart rate for a given work rate will provide feedback on the benefits of heat acclimatization. Similarly the use of Ratings of Perceived Exertion is likely to show the athletes that they perceive a given level of exercise to be less strenuous after heat acclimatization. The monitoring of the above during exercise in the heat acclimatization period is likely to promote confidence in the athletes who will have evidence that they are able to cope better when they exercise in heat. They will experience less heat distress and discomfort and their cognitive function will be improved. Recording of the exercise bout details (e.g. work rate, and duration) will highlight to the athletes that they are able to do more and complete sessions that were not possible at the start of the acclimatization period.

Sweat-inducing clothing will create a wet internal environment and simulate a tropical heat in cool conditions. This strategy has been shown to be effective in reducing core temperature in heat, but the benefits were not so great as training in a hot environment. It is better to carry out the heat acclimation training sessions in a botanical gardens or heat chamber. Normal training sessions should be maintained before the heat-acclimatization sessions or at other times of the day. Nearer the time of the event, the exerciser should travel to the area where the event will take place. Training sessions can be conducted during the cool parts of the day or indoors in an air-conditioned room. Heat-acclimatization sessions can be conducted in the heat of the day to gain benefits from heat stress.

Timing

Heat-acclimatization changes are fairly fast. Lind and Bass (1963) showed that changes in heart rate, sweat rate and core temperature started to plateau in about a week. However, the physiological adaptations may not be parallelled by changes in performance. Montain, Maughan & Sawka (1996) recommend that elite athletes should have 1 month of heat acclimatization before an important event. However, the above adaptations show that heat acclimatization of even short duration provides advantages for the exerciser in heat.

There is individual variation in response to heat acclimatization, which cannot be predicted from any marker. Individuals who do not tolerate heat particularly well, or who take longer than usual to adapt, should be identified. Monitoring of exercise heart rate and Ratings of Perceived Exertion from before heat acclimatization and during heat acclimatization may help identify those who are slow to acclimatize to the heat. Provision may then be made for such individuals to have a longer acclimatization period than normal.

It is not necessary to carry out heat-acclimatization sessions on consecutive days. Benefits are likely to be similar using 10 consecutive days (10 calendar days) and 10 sessions every third day (27 calendar days).

Early research suggested that heat acclimatization of 100 minutes per session was desirable to gain any benefits. More recent studies have suggested, however, that heat-acclimatization benefits can be gained from fairly short sessions. Comparison of 60-minute sessions at 50 per cent of the maximal oxygen uptake (VO_2max), and 30- to 35-minute sessions at 75 per cent of the VO_2max, showed that there was a similar decrease in core temperature and heart rate in a 90-minute heat-tolerance test at 50 per cent of the VO_2max (Houmard et al., 1990).

A one-month period allows a flexible training regime to be structured. Initial acclimation at home allows the athlete to gradually introduce 'heat' sessions without the added stressors of travel fatigue, time zone changes and disruption to normal routines. Travel 14 days before an event will allow the athlete to experience the heat of the new environment, to appreciate how hot it is, to have time to adopt new routines to cope with the heat, and to build the athlete's confidence so that he can cope with the conditions. In addition, a 14-day period before competition will allow the athlete to adapt to the new time zone. Finances and time availability will, of course, have to be taken into account when structuring a month of heat acclimatization.

DECAY (LOSS OF HEAT-ACCLIMATIZATION BENEFITS)

The loss of heat-acclimatization benefits (decay) would appear to be fairly fast. Even after a week of removal from hot conditions there may be some loss of heat-acclimatization benefits. Studies report a range of times for decay, which may be due to individual variability and different study designs. After a week with no heat stress there was a loss of benefits: 25 per cent in core temperature, a 50 per cent loss in heart rate and a 50 per cent reduction in sweat rate (Williams et al., 1967). Lind and Bass (1963) suggest that the beneficial effects of heat acclimatization may be lost at different times depending on the marker used. They indicate that rectal-temperature benefits may be maintained longer than reduced heart rates resulting from heat acclimatization.

High fitness levels and warm weather may enhance retention of heat-acclimatization benefits.

Heat-acclimatization benefits are lost more rapidly if the exerciser becomes dehydrated. Sawka et al. (1983) showed that core temperature was lower after heat acclimatization. However, when the heat-acclimatized subjects were dehydrated, core temperature increased. The core temperatures were very similar when the heat-acclimatized and non-heat-acclimatized subjects were both dehydrated. Thus, dehydration had a greater impact on the heat-acclimatized subjects. The larger increase in core temperature when they were dehydrated shows that they were penalized more than the non-heat-acclimatized subjects were. Heat-acclimatized subjects have more to lose when they become dehydrated. The heat-acclimatized athlete also sweats more, thereby needing an increased fluid intake.

Loss of heat-acclimatization benefits

- Heat-acclimatized subjects have more to lose when they become dehydrated.
- Their core temperature increase as a result of dehydration is greater than that of non-heat-acclimatized subjects.
- Skin disorders (rash, sunburn) limit sweating and thus diminish heat-acclimatization benefits relating to sweating.
- There is some evidence that sleep deprivation and illness reduce heat-acclimatization benefits.

How to achieve heat acclimatization

- Mimic the competitive environment.
- Expose the body to repeated heat stress that increases core temperature and sweating.
- On arrival in a hot environment, the athlete should start slowly, and avoid any possibility of heat injury.
- Identify appropriate hot and cool times for heat acclimatization and quality training sessions.
- Train in the cool, using high-intensity sessions
- Acclimatize in the hottest part of the day to gain benefit from heat stress.
- Avoid a loss of heat-acclimatization benefits by preventing dehydration and sunburn.
- Replace water loss.

RE-ACCLIMATIZATION

While heat-acclimatization benefits are lost with removal from repeated heat stress, they can be regained fairly fast. Core temperature was reduced from 37.7°C to 37.0°C after 8 days of heat acclimatization. After 21 days of no heat stress and a day of heat re-acclimatization, core temperature had increased to 37.3°C, giving only 43 per cent of the heat-acclimatization benefit after 8 days of heat acclimatization. On Day 3 of re-acclimatization, core temperature had decreased to 37.0°C, the same value as after 8 days of heat acclimatization.

The picture for heart rate is similar. Heart rate decreased from 177 to 151 beats/min after 8 days of heat acclimatization. On Day 1 of re-acclimatization,

21 days after the cessation of the 8 days of heat acclimatization, the heart rate had increased to 165 beats/min, but by Day 3 was 150 beats/min, slightly lower than after 8 days of heat acclimatization. Thus, despite 21 days removed from heat stress, the subjects were able to regain their physiological benefits from heat acclimatization after only 3 days of exposure to heat (Pichan et al., 1985).

CHAPTER 5

Exercise in Cold

A SURVIVOR OF EXTREME COLD

Skier Anna Bagenholm was exposed to cold and wet conditions resulting in a core temperature of 13.7°C, but she survived.

Anna's enjoyable day on skis came to an abrupt end when she found herself trapped in a waterfall. Her friends were unable to extricate her from her jammed position, but an air pocket under the ice enabled her to survive, despite her core temperature plummeting. She remained under the ice with cold water pouring over her until her friends finally pulled her out, 80 minutes after she was first trapped. During her helicopter flight to hospital she was given cardiopulmonary resuscitation. In hospital she was warmed up using cardiopulmonary bypass (a technique that circulates the blood through a machine outside the body, where it is warmed before being returned to the patient). Anna spent 60 days in intensive care before beginning her rehabilitation. Today, she has recovered completely, apart from a tingling sensation in her hands.

The story of Anna's recovery is a remarkable one, since her core temperature fell to 13.7°C, dramatically lower than the normal core temperature of 37°C. If the core temperature falls to less than 30°C, most cold casualties die. Before Anna's case, the lowest recorded core temperature of a human who survived was 14.4°C. There are other examples of individuals who have been immersed in cold for many minutes who have been resuscitated.

When the human body is cooled, the heart rate is slowed and oxygen consumption is decreased. The brain can survive with much less oxygen than normal. Thus, while a 'cold' person may look lifeless (it may be difficult to detect breathing or a heart rate), it may still be possible to resuscitate him. As the maxim states, 'a person is not dead until he is warm and dead'. (*See* Gilbert et al. (2000), Resuscitation from accidental hypothermia of 13.7°C with circulatory arrest, *The Lancet*, 355, 375–6.)

Effects of extreme cold and some warnings

- The body cools rapidly when exposed to cold and wet conditions.
- It is essential to treat cold casualties gently as movement can trigger cardiac arrest (heart stoppage).
- It should not be assumed that apparently lifeless cold individuals are dead. Resuscitation, sometimes prolonged, has been effective in cold individuals who were apparently dead.

EXERCISE IN COLD

Endurance performance has been reduced by 38 per cent (119.9 mins to 66.9 mins) in –20°C compared with +20°C (Patton and Vogel, 1984). Scandinavian cross-country ski race times have also been shown to be slower in very cold conditions, at around –28°C compared with the 'normal' –7° to –12°C. The combination of frostbite and cold discomfort forced some participants to drop out.

Some marathoners, who have competed where the mean ambient temperature was below zero, have been thought to have a reduced performance as a result of the cold conditions. It has been speculated that they may have experienced a bronchoconstriction (narrowing of the airways as in asthma). Subjects do not have a reduced maximal oxygen uptake (VO_2max) in acute exposure to cold air at least to temperatures as low as –20°C.

Marathon running is often associated with hyperthermia, but 'slow' runners may be unable to generate sufficient heat to offset the cooling effect of wind and rain, and some runners have below normal core temperatures at the end of a marathon. Athletes involved in long-distance events, including marathons, have experienced problems in what might appear to be fairly benign conditions.

Some triathlons take place in early summer where open-water swims may take place in water temperatures lower than 17°C. In a triathlon with a water temperature of 15.2°C and no wind, the winner of the 1-km swim took 15 minutes. Subjects who took longer than 25 minutes for the swim required assistance from the water if they were wearing less than a short-legged suit. One lean marathon runner was not able to continue after the swim (Lowdon, 1984).

PHYSIOLOGICAL RESPONSES TO COLD

When an individual is exposed to cold, there is a reduced blood flow to the extremities and the skin. This vasoconstriction (narrowing) of skin blood vessels, and of those to the extremities, reduces the flow of blood and therefore the flow of heat to the skin. The skin temperature is lowered, the temperature gradient between the skin and the environment decreases, and this in turn reduces heat loss. It should be noted that there is limited vasoconstriction in the scalp, which highlights the need to protect the head in the cold to limit heat loss.

Exercise responses to cold

During exercise in the cold, there is reduced vasoconstriction compared with rest. The working muscles generate heat and require an enhanced blood flow to supply oxygen to the muscles and to carry away waste products. Thus, areas around the working muscles have a higher temperature compared with resting muscles. The increased blood flow and temperature reduces the tissue insulation and thus facilitates heat loss.

Vasoconstriction may not be sufficient to limit heat loss by itself. The person may be unable or unwilling to exercise and therefore is unable to generate heat through exercise. In this case, further cooling will take place. It may not be possible to put on more clothes. Shivering – the involuntary contraction of a muscle group against its antagonists, for example, quadriceps against hamstrings – is then triggered to increase heat production. Shivering can increase heat production 4–5 times above the normal resting level. However, shivering not only impairs skilled movements, it also enhances heat loss in areas surrounding the shivering muscles, as a result of increased blood flow to the shivering sites. Shivering increases the energy cost of an activity and may increase carbohydrate utilization and result in premature fatigue. It is a sobering thought that in most cold conditions it is not physiology that has allowed man to survive but behavioural responses, such as use of clothing, taking shelter (including housing), and use of fire (and its successors).

COLD AND PERFORMANCE

Comparing Normothermia and Low Core Temperature

Davies et al. (1975) compared normothermia and low core temperature (LCT) using submaximal and maximal exercise. They sat their LCT subjects in cold water until their core temperature fell to 35°C. The 'normothermia' subjects had a 'normal' core temperature. In submaximal exercise (100 watts), VO_2 was higher in LCT because of shivering (*see* Table 4). VO_2max was lower in LCT compared with normothermia. These results show that subjects with LCT are less efficient submaximally – they use more O_2 than in normothermia and they have a decreased maximum.

Table 4 **Low core temperature and exercise**

	Air breathed out per min (VE) (L/min)	*Oxygen used (VO_2) (L/min)*	*Heart rate (beats/ min)*	*Work rate (watts)*
Normothermia	41.1	1.52	114	100
Low core temperatures	64.1	1.90	100	100
Normothermia	165.0	3.87	184	MAX
Low core temperatures	145.3	3.45	158	MAX

(Adapted from Table 1 in Davies et al. (1975), The effects of hypothermia on submaximal and maximal work performance, *Acta Physiol Scand*, 95, 201–2. Used with permission from the Scandinavian Physiological Society.)

The implications of a low core temperature, based on the Davies et al. (1975) study, are as follows:

- low core temperatures increase the energy cost of an activity during submaximal exercise and reduce VO_2max;
- athletes with low core temperatures are therefore less efficient submaximally and have a reduced work capacity.

Cold Versus Hot Conditions

Cool conditions have been shown to improve endurance performance. During exercise at 70 per cent VO_2max at four different air temperatures – 4, 11, 20 and 31°C – endurance time was 94 minutes at 11 °C compared with 52 minutes at 31°C with endurance times of 82 minutes for the other two temperatures (Galloway and Maughan, 1997).

HOW COLD IS COLD?

When estimating the severity of cold stress, consideration must be given to wind and degree of wetness. Wind (convective heat loss) has a significant cooling effect on the body (*see* Table 5), and a body is losing more heat at 4°C in a 30mph wind than at –1°C in still air. (The wind-chill equivalent temperature for any wind speed is the temperature at which the person would be losing the same amount of heat without any wind; *see* Table 5.) Note that, as the wind speed increases, there is not a proportional cooling effect. At wind speeds over 30mph, there is a much lower increase in cooling effect compared with lower wind speeds.

Table 5 **Wind-chill chart**

	Actual thermometer reading (°C)				
	4	*–1*	*–7*	*–12*	*–18*
Wind speed (mph)	*Equivalent temperature (°C)*				
Calm	4	–1	–7	–12	–18
10	–1	–9	–15	–23	–29
20	–7	–15	–23	–32	–37
30	–12	–18	–29	–34	–46
40	–12	–21	–29	–37	–48

The Wind Chill Index shows the cooling effect of wind on the body, and indicates that the cooling effect is greater with an increased wind speed. (Adapted from Lloyd, E. (1994), ABC Sports Medicine, *British Medical Journal*, 309, 531–4. Reproduced with permission from the BMJ Publishing Group.)

The cooling effect of wind is affected by the movement speed and direction of the exerciser. For example, a runner running at 10 mph, in an ambient temperature of –1°C, with a wind speed of 20mph, has an effective wind speed of 30mph if running into the wind (20mph + 10mph) giving a wind chill of –18°C. However, if running with the wind, the effective wind speed is 10mph (20mph – 10mph) giving a wind chill of –9°C.

OTHER FACTORS AFFECTING RESPONSES TO COLD

Skinfold Thickness

A thick layer of fat is an advantage for an individual who is in a cold environment as the fat provides insulation. Lean individuals are more vulnerable to cold than fat individuals. The head is a part of the body with limited subcutaneous fat and this increases the danger of heat loss from the head.

Body Surface Area

Heat has to be generated to maintain core temperature but heat is lost through the body surface. Therefore the larger the surface area to body mass ratio, the more vulnerable the individual is to heat loss. Children are more susceptible to cold than adults since they have more body surface area per unit weight; a 7-pound baby, for example, has 2.5 times that of a 140-pound adult. An exposed head is a major source of heat loss in an adult because of lack of subcutaneous fat and poor vasoconstrictor capability. In a young child it is an even more important source of heat loss since it makes up a larger proportion of the body, and, therefore, of the body surface. In neonates, insulating the head and leaving the rest of the body naked is more effective at maintaining normothermia than insulating the body and leaving the head uncovered.

Aerobic Fitness

Around 25 per cent of the energy from muscular contraction is used for mechanical work, with about 75 per cent of the energy given off as heat. This 75 per cent usually results in an increased core temperature. In cold environments, exercise at low intensity may not be sufficiently high to generate sufficient heat to offset the environmental cooling effects. During moderate and high-intensity exercise, sufficient heat may be generated to maintain core temperature. Thus, the individual who can sustain high-intensity exercise for long periods of time may be able to prevent a fall in core temperature, whereas the unfit person may need to slow down or stop, resulting in a decreased heat generation and a reduced core temperature. (*See* Table 6 in this volume.)

The VO_2 was higher in Situation B during low-intensity exercise, but there was no difference in the oxygen used (VO_2) at high intensities. The higher oxygen used (VO_2) at low and moderate intensities in Situation B was due to shivering, which resulted from the need to generate heat as a consequence of the cooling effect of the convective heat loss and the reduced insulation of the wet clothing. However, shivering was not necessary to generate heat at high intensities, as the heat generation was great enough to offset the cooling effect of the Situation B environment.

Table 6 The effect of wind and wet clothing on core temperature during exercise

Work (kg/m/min)	Oxygen used (VO_2) (L/min) A	Oxygen used (VO_2) (L/min) B	Rectal temp (°C) A	Rectal temp (°C) B
REST	0.28	1.02	37.19	36.57
400	1.06	1.60	37.55	36.51
600	1.60	2.00	37.93	37.93
900	2.19	2.18	38.13	37.77
1200	3.06	2.91	–	–

Situation A: Dry, hill-walking clothing. No wind.
Situation B: Wet clothing. Wind speed 15km/h.

(From Table 2, page 334, in Pugh, G. (1967), Cold stress and muscular exercise, with special reference to accidental hypothermia, *British Medical Journal*, 2, 333–37. Reproduced with permission from the BMJ Publishing Group.)

Wetness

Heat conductance is 25 times greater in water than in air. Thus, the individual who is wet is much more vulnerable to cooling than a dry individual. A wet environment – cold water or wet clothing – that results in a reduction in the insulation of the clothing promotes greater heat loss compared with a dry environment.

Cold and Exercise Intensity

Athletes who exercise in the cold at a low intensity as a result of fatigue, injury or to accompany slow athletes, are susceptible to an increase in VO_2, as a result of shivering.

Any athlete who is unable to sustain fairly high exercise intensity, due to lack of fitness or injury, in cold conditions, must have appropriate clothing or suitable shelter, if they are to avoid body cooling.

CLOTHING

During training or in competition the athlete may not need much clothing, as he is able to generate sufficient heat to combat the normal environmental heat loss. However, in training sessions and in competition, conditions may vary greatly (for example, there may be a great difference in the cooling effect between running into a head wind and running with a tail wind), and adjustment of clothing is desirable. Athletes will rest in cold conditions and must also make clothing adjustments.

Layers

Clothing for cold environments should consist of layers:

- The *under layer* should provide warmth and have the ability to wick moisture away from the skin. The removal of moisture is important as water conducts heat 25 times faster than air.
- The *insulation layer* may need to be adjusted. The prevailing environmental conditions will determine how much insulation is needed. As exercise intensity will vary and the weather conditions will change, there will be a need to remove or add clothing, and/or open or close zips. Some materials, such as down, may offer a great deal of insulation, but these can be bulky. The weight of the insulation is also a consideration. Heavy or bulky clothing may increase the energy cost of an activity.
- The outer layer, or *protection layer*, should offer protection from wind and rain and have the capacity to allow water vapour (resulting from sweating) to pass through it to the atmosphere. The latter characteristic may be difficult to accomplish completely when the individual is exercising at high intensity which results in high sweat rates and a considerable build-up of moisture.
- Attention must also be paid to *air layers and seals.* Air is a poor conductor of heat. Movement can produce a bellows effect, where warm air is squeezed out and cold air drawn in, in a rhythmical manner. Clothing should be arranged to avoid unnecessary air circulation and seals around the ankles, wrists or neck must be adequate to prevent this. Seals must have draw cords, which constrict clothing and prevent air movement. Gloves and sleeves should overlap at the wrist. Salopettes will prevent penetration of air at the waist, which can be a problem with bibless trousers. Clearly, an outer layer that is not windproof will facilitate air circulation.

Clothing Adjustment

It is not enough to have clothing that has the characteristics of the three layers outlined above. The exerciser should always attempt to stay dry, although it is often difficult to remain dry from within, as heavy exercise results in sweating and moisture build-up. Heat and moisture production can be controlled by regulating the exercise intensity and by adjusting clothing to avoid unnecessary cooling or overheating. Reduction in exercise intensity to avoid heavy sweating is not realistic during competition, but the strategy may be appropriate in training situations. Where it is desirable to prevent overheating, areas that

Adjustment of clothing

It is not enough to have appropriate clothing. The exerciser must also be able to make adjustments to allow for changes in weather conditions and exercise intensity. The aim is to avoid unnecessary build-up of moisture during exercise, and to limit heat loss while resting, or running or skiing into a head wind. The head should be covered when resting.

Fig 13. Layers and adjustment to clothing during exercise.

A hood that extends well beyond the face, to give protection; the outer windproof layer partially open to provide some ventilation; the first insulation layer zipped up to limit air movement. Mitts keep the hands warmer than gloves. (Reproduced with permission from Mr Ian Donnelly)

promote heat loss (such as the head) should be exposed to the air, but should be covered when cooling is a possible problem. Removing clothing and/or opening zips and vents are possible strategies to prevent moisture build-up. Opening of zips in areas such as the armpits and groin can reduce heat and moisture build-up.

Overdressing results in unnecessary sweat production and unnecessary wetting of clothes. An example of inappropriate defences against cold, and of the importance of clothing adjustment, is the overdressed soldier who was shovelling snow on a cold Alaskan day. He was treated for heatstroke in hospital.

If the exerciser stops to rest, the skeletal muscles have reduced much of their heat generation, and it is necessary to adjust clothing immediately, by closing zips and putting on clothes to prevent heat loss.

Headgear

Blood vessels in the head do not constrict in response to cold compared with other areas in the body. Thus, the head is a potent site of heat loss, particularly in bald individuals or young children. In some conditions, a balaclava that covers almost all of the head and neck and a hood may be necessary. The hood of the outer layer jacket should have a stiff malleable wire around its edge so that the face is afforded some protection, and the hood is not disturbed by high winds. A mask and scarf around the mouth and nose (to avoid frostbite in the nose) may also be necessary in very harsh conditions while a hat or ear muffs may suffice on other occasions.

Protection for the Hands

Mitts will provide better protection against the cold than gloves, but they do lead to a loss of dexterity. The use of thin gloves may be of value when manipulative skills are necessary; they will prevent direct skin contact with, for example, metal, and the possible removal of skin in very cold conditions. Gloves also offer protection from the wind. The combination of fingerless gloves under full gloves or mitts allows full finger dexterity to be used for short intervals with minimal heat loss. Some mitts have a fold-back pouch that may be used to enclose the fingers. This has the same benefits, with the advantage that the finger pouch is attached and will not be lost.

Lower Body

A close-fitting synthetic fibre underlayer and waterproof/windproof trousers may be needed in some conditions.

The combination of thin polypropylene socks next to the skin and woollen socks as a second layer will help remove moisture from the skin. Footwear and socks should be dried whenever possible.

Clothing Material

It is important to select an appropriate material when considering clothing. For example, the way a given material reacts to moisture may influence how it feels and its insulative capability. Different materials have different properties:

- *wool* is a good insulator, with limited loss of insulation when wet;
- *cotton* takes in moisture easily and has great loss of insulation when wet (up to 90 per cent);
- *polypropylene* is good at transferring moisture and is often used for thermal underwear;
- *nylon* offers good insulation, dries quickly, but offers limited defence against water and wind;
- *polyester*, often used in fleece jackets, is a good insulator;
- *down* gives excellent insulation, but is bulky and there is a great loss of insulation when it is wet.

(Adapted from Dickinson, 1995)

DANGERS OF A COLD ENVIRONMENT

Exposure to cold may cause many problems, which may be grouped together as follows:

- local effects;
- general effects (affecting the whole body, with a marked drop of core temperature); and

- secondary effects (which may occur with only a slight or no fall of core temperature).

Local Effects of Cold

Exposure to sub-zero temperatures may lead to frostnip or frostbite.

Frostnip

In frostnip the exposed skin becomes painful, then blanches and loses sensation, but remains pliable. The part should be warmed by placing in the armpit or under clothing. The part should tingle, have an increased blood flow (hyperaemia), and within a few minutes sensation should be restored and normal activity can be resumed.

Frostbite

Frostbite is likely to limit performance in skilled movements and can lead to a decrease in manipulative skills and a loss of body parts. Frostbite is a localized lesion caused by freezing. It preferentially affects the periphery – feet, hands, ears, nose and cheeks – although the cornea has also been affected in downhill skiers and in speed skaters unprotected by goggles. Penile freezing can occur during skiing or running in tight or inadequate clothing or from direct contact with a metal zip. Anything that restricts the circulation, such as tight training shoes or climbing boots, increases the risk, as does dehydration.

In frostbite the tissues are hard, insensitive and white or mottled (*see* Figure 14). No attempt should be made to thaw frostbite if there is any chance of the part becoming refrozen because freeze-thaw-refreeze causes more damage than continuous freezing. It is safer to continue to walk on frozen feet even for as long as 72 hours.

Treatment involves rewarming that is ideally done in a hot (40°C) whirlpool bath, with gradual spontaneous rewarming as second best. Beating, rubbing with snow or rewarming with excessive heat produce disastrous loss of tissue.

Because damage from frostbite is usually more superficial than first suspected, amputation should be delayed for up to 90 days till mummification and demarcation are complete. A few hospitals with special expertise have specialized investigation methods – for example, the infusion of technesium 90 – which facilitates earlier decisions. After recovery the sufferer can return to full activity, limited only by any tissue loss.

Trench Foot

Trench foot is the usual title for non-freezing cold injury, which develops over a fairly long period when the feet are exposed to cold (above freezing) and/or wet, as they are during prolonged walking in boggy ground or when sweating in impervious boots. As in frostbite, damage is more likely if there is also fatigue, dehydration, immobility and tight footwear. The feet are initially cold and numb, giving the sensation that has been described as being like 'walking on cotton wool', and the combination with joint stiffness causes the victim to walk with legs apart to maintain balance.

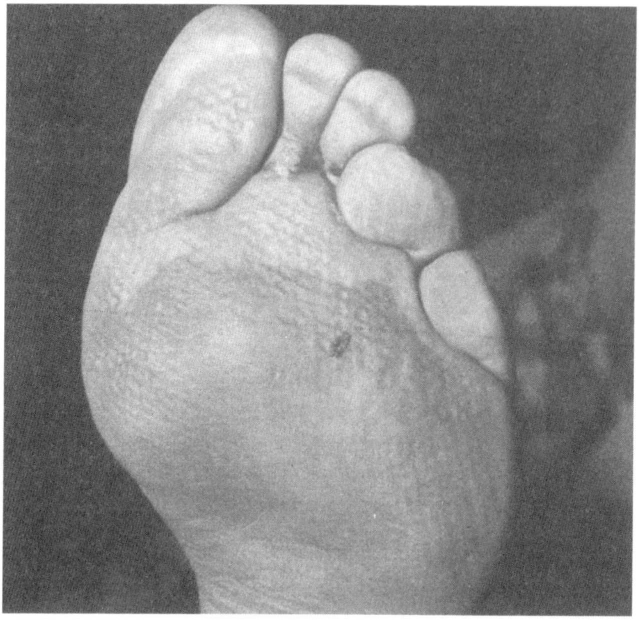

Fig 14, A frostbitten foot – cold and rigid, without sensation or digital motion.

Note marks of socks.

(Photo: Dr W. J. Mills Jr of Anchorage, Alaska, previously published in *Alaska Medicine*, 1993, vol 35, page 46 and reproduced with permission)

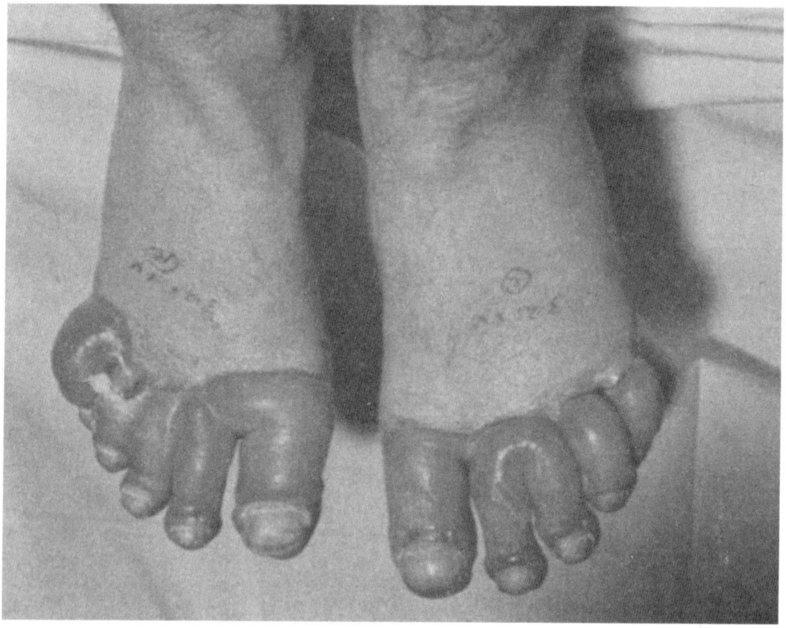

Fig 15. A frostbitten foot after rewarming, with blisters.

The clear blisters become black when dry but are finally sloughed off, leaving normal anatomy and function. Blood-filled blisters indicate greater tissue damage. (Photo: Dr W. J. Mills Jr of Anchorage, Alaska, previously published in *Alaska Medicine*, 1993, vol 35, page 47 and reproduced with permission)

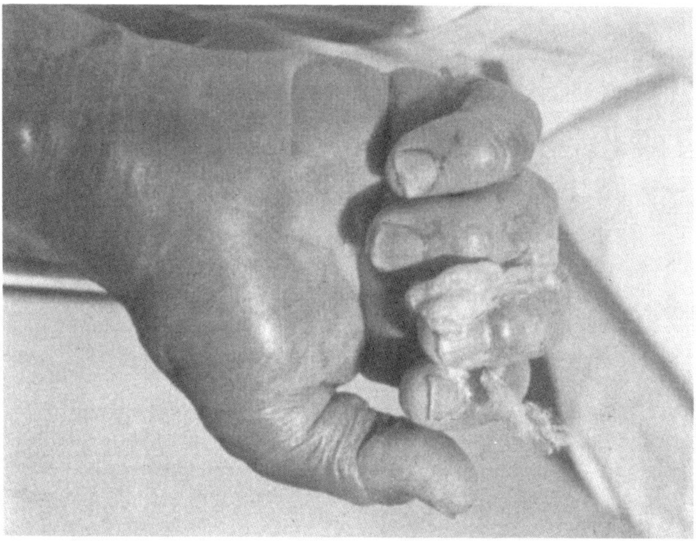

Fig 16. Frostbitten hand thawed by pouring scalding water over it.

Fewer than 24 hours after injury, the hand had become cyanotic, quite painful and foul-smelling. There was no formation of blisters. The frostbite had been converted to a severe burn. (Photo: Dr W. J. Mills Jr of Anchorage, Alaska, previously published in *Alaska Medicine*, 1993, vol 35, page 43, and reproduced with permission)

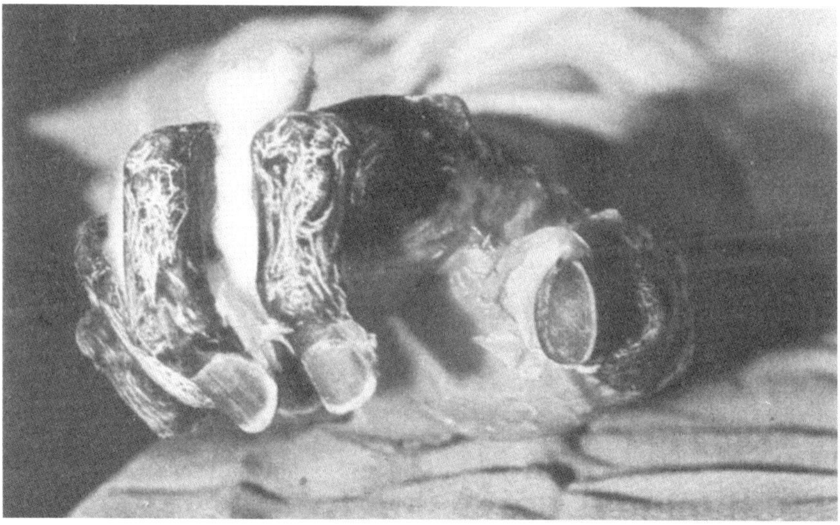

Fig 17. Three weeks after being frostbitten, the digits on this hand are hard and rigid, with the soft tissue mummified.

There is clear demarcation between the tissue that is dead and the tissue that is still alive. Spontaneous amputation of all the fingers of both hands occurred after six weeks. (Photo: Dr W. J. Mills Jr of Anchorage, Alaska, previously published in *Alaska Medicine*, 1993, vol 35, page 43, and reproduced with permission)

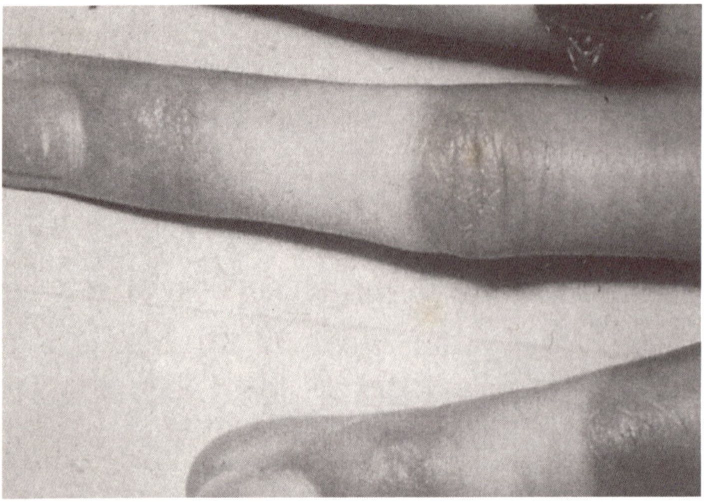

Fig 18. Raynaud's phenomenon in two fingers.
(Photo: E. Lloyd)

When first seen, the feet are cold, swollen and blotchy pink-purple, or blanched. After rewarming they become hyperaemic (increased blood flow), hot and red with paraesthesia (tingling sensation) and/or pain, which may be severe, like electric shocks, especially on weight bearing. This may last for weeks. Severe cases have bleeding into the skin, ulceration, and/or blisters and may progress to gangrene.

The individual is treated by removal from the hostile environment and the part is allowed to rewarm spontaneously. Warm-bath rewarming is only indicated if the pulses are impalpable.

Trench foot is likely to limit performance and can lead to a permanent problem with the feet. Long-term effects are likely to be persistent hypersensitivity to cold as well as anaesthesia or hyperaesthesia (oversensitivity, or pain on light touch), and problems with the bony structures of the feet.

Trench foot can be prevented by regular inspection and avoiding having continuously damp feet.

Raynaud's Syndrome

Some individuals respond to cold stress with a very severe arterial vasoconstriction, most commonly affecting the fingers (*see* Figure 18). In severe cases this may lead to digital ulceration and tissue loss. Protection from cold by suitable clothing, gloves and shoes is usually sufficient, but chemical hand warmers, or electrically heated gloves or shoes may be necessary in extreme cases.

Chilblains

Chilblains are local inflammatory lesions that develop as a result of exposure to cold. The skin becomes brittle, which leads to chapping and itching. They are more common in a humid climate and inadequate clothing is an important predisposing factor. Central heating has reduced the incidence (Lahti, 1982). Chilblains usually occur on feet but are also liable to occur on the insides of the thighs in horse riders.

Winter Itch

A low relative humidity can result in dry, itchy skin, which is made worse by central heating if the temperature is kept high. Treatment involves humidification of the room, emulsifying skin ointments and the avoidance of too much washing (Lahti, 1982).

General Effects of Cold

Deterioration in Performance

See page 72.

Hypothermia

Hypothermia, defined as occurring when the body's core temperature falls below 35°C (normal 37°C), or sees a decrease of 2°C. There are four physiologically distinct types of hypothermia (*see* below) and inappropriate treatment may lead to death. Suitable conditions may occur in many activities and at any time of year.

The four types of hypothermia are as follows:

1 Submersion: the person is totally submerged (including the head) in ice-cold water. The rate of cooling is very rapid (up to 30°C per hour), partly because vasoconstriction is not efficient in the head. Survival is possible after prolonged submersion (up to 90 minutes), and is more likely in children.
2 Immersion: the person is in the water but with the head out. This is very severe cold stress – core temperature falls despite maximal heat production. The commonest cause is immersion in cold water.
3 Exhaustion: this is less severe cold stress – heat production can maintain core temperature until exhaustion occurs. It is most commonly found in mountaineers or hill walkers but can also occur in runners and cyclists in long-distance events.
4 Urban: the cold is relatively mild but prolonged – core temperature remains normal (35°C or above), possibly for weeks, before hypothermia develops. It is most usual in the elderly or in cases of malnutrition.

Immersion and exhaustion hypothermia are the types usually found in sport and exercise, although submersion hypothermia is also possible.

Diagnosis is crucial. The earliest symptom is usually a change in behaviour of the person, but similar symptoms occur in hyperthermia, exhaustion, hypoglycaemia (low blood-sugar levels) and even a mild cerebral stroke. This is followed by uncoordinated movements, staggering, a progressive clouding and loss of consciousness, and slowing of heart and respiration, with death as the final outcome.

For some symptoms and signs of hypothermia *see* Table 7, but these are an unreliable guide to core temperature. For example, a boy rescued from cold water was conscious and shivering from the moment of rescue, although his core temperature was calculated as being 27°C when first rescued. Similarly an 84-year-old lady was semi-conscious but responding to speech with a core temperature of 24.3°C. By contrast, mountain rescue teams often report casualties with core temperatures about 35°C who are not shivering. (There are also many causes of

Table 7 **Signs and symptoms at different levels of hypothermia**

Core temperature in °C	*Signs and symptoms*
37.6	'Normal' (rectal temperature)
37	'Normal' (oral temperature)
36	Increased metabolic rate to attempt to balance heat loss
35	Shivering is maximum
34	Patients usually responsive and with normal blood pressure; lower limit compatible with continued exercise
33–31	Retrograde amnesia, consciousness clouded, blood pressure difficult to obtain, pupils dilated, most shivering ceases
30–28	Progressive loss of consciousness, increased muscular rigidity, slow pulse and respiration, cardiac arrhythmia develops, ventricular fibrillation may develop if heart irritated
27	Voluntary motion lost along with pupillary light reflex, deep tendon and skin reflexes. Subject appears dead
26	Victim seldom conscious
25	Ventricular fibrillation may appear spontaneously
24–21	Pulmonary oedema develops. 100 per cent mortality in shipwreck victims in Second World War
20	Heart standstill
18	Adult *accidental* hypothermia with recovery by spontaneous rewarming (Laufmann, 1951)
16	Adult *accidental* hypothermia with recovery (Wood, 1977). Lowest infant *accidental* hypothermic patient with recovery (Nozaki et al., 1986). Adult *accidental* hypothermia with recovery (Burgess, J., Personal Communication)
13.3	Lowest adult *accidental* hypothermic patient with recovery (Gilbert et al., 2000).
9	Lowest artificially cooled hypothermic patient with recovery (Niazi & Lewis, 1958)
4	Monkeys revived successfully (Niazi & Lewis, 1957)
1 to –7	Rats and hamsters revived successfully (Smith, 1959)

(Adapted from Harnett et al., 1983.)

loss of consciousness with normothermia.) Accurate diagnosis can only be made by measuring the core temperature, ideally oesophageal but usually rectal, and the necessary low-reading thermometer may not be available. For practical purposes a subject should be treated as a 'cold casualty' if the body feels 'as cold as marble' and, in particular, if the armpit is profoundly cold.

With all types of hypothermia, further heat loss must be prevented by removing the casualty from the cold environment. Movement must be gentle to avoid triggering a cardiac arrest, but in a severe environment such as cold water, the priority is to get the person out. Until the casualty is indoors, wet clothes should not be removed, but layers of insulating material should be put on top of the clothing and covered with a layer which is water- and wind-proof. The head must be included. A 'space blanket' is often recommended but it is no more effective than a piece of polythene of a similar thickness, which is very much cheaper.

The above measures will probably be sufficient before admission to hospital in a case of immersion hypothermia. If available, airway warming (the inhalation of warmed moist air) should be used. This treatment is of particular value in exhaustion hypothermia where it will prevent the sometimes fatal cardiovascular collapse that may occur during rewarming. Moderate warmth is dangerous in exhaustion hypothermia. In a case of urban hypothermia, *no* additional heat, either surface or central, should be supplied before admission to hospital as this may precipitate fatal pulmonary (lung) or cerebral (brain) oedema (fluid build-up and swelling) or both, through reversing intercompartmental fluid shifts. These shifts can be monitored and controlled in intensive care. With submersion hypothermia, resuscitation must start immediately on rescue.

Ideally, all hypothermic casualties should be taken to hospital, where rewarming should be done with intensive-care monitoring.

A cold casualty can sometimes be asked to continue to walk, but this carries the risk of complete exhaustion, which can result in a precipitous fall in core temperature. Exercise should only be attempted if the person is not exhausted, and if the party is within reach of safety, or if circumstances are such that it is impossible to carry the casualty – there may be too few in the party, or the risk of leaving the person and returning with a rescue party may be considered too great.

A fairly hot bath is the traditional method of rewarming but it is seldom available outside major habitation. It is also impossible to monitor the casualty or carry out resuscitation in the bath.

Resuscitation in hypothermia should be carried out as follows:

1 respiratory obstruction should be cleared and, if necessary, expired air ventilation started using the same criteria and rate as in normothermia;
2 cardiopulmonary resuscitation (CPR), at the same rate as in normothermia, should be started if indicated:
   (a) if no carotid (neck) pulse is detectable after feeling in the correct place for at least one minute; *or*
   (b) if cardiac arrest is observed (a pulse disappears); *or*
   (c) if there is a reasonable chance that cardiac arrest occurred within the previous two hours; *and*
   (d) if there is a reasonable expectation that effective CPR can be continued with only brief periods of interruption for movement, until the casualty can be transported to hospital, where full advanced life support can be provided.

NB: Casualties who have been known to be totally submerged in very cold water can recover even after submersion of up to 90 minutes, especially if they are young.

Secondary Effects of Cold

Loss of Manual Dexterity

Individuals who carry out tasks with their fingers can expect a loss of dexterity and a decrease in sensitivity when their hands or forearms are cooled. Those who regularly need to carry out intricate skilled movements – sailors tying knots or biathletes trying to load their rifle in a hurry, for example – should be aware that cooling of the hands and fingers can cause major decrements in manipulative skills. A sailor may drop a vital piece of equipment, which then slides overboard. The effect will be aggravated by shivering, and by frostbite, which has the other danger of loss of body parts.

Decrements in skilled performance are more marked when cooling is fast compared with slow cooling. With slow cooling there is also a performance loss after rewarming. Factors that may contribute to the performance decrement include the need to wear bulky mitts or gloves for cold protection and diminished flexibility in the forearm and finger muscles.

Muscle Impairment

Cold-induced vasoconstriction (narrowing of the blood vessels) reduces the blood supply to the muscles, and the muscles themselves are therefore cold. Muscle strength is decreased when muscle temperature is lowered. This may be because the cold impairs the functioning of the enzyme systems. Cooling of the forearms has been shown to reduce hand-grip strength and endurance. There is also diminished muscular flexibility.

Muscle Injury

Muscles are inefficient when cold – they do not contract as smoothly as they do when they are warm. The fluid in the joints is more viscous when cold so that more effort is required to move the joint. Nerve conduction is slowed by cold so there may be slight impairment of coordination of the contraction of different groups of muscles or even within an individual muscle group. Muscle and tendon tears therefore can occur when a person is cold.

In order to prevent such problems, the warm-up must be sufficiently energetic and prolonged, until the limbs feel 'glowing warm'. The whole body should feel warm, including the limb muscles, the musculotendinous junction (the join between a muscle and its tendon), the muscle insertion (where it is attached to a bone), and the joints. It is tempting to try and use a hot shower as a warm-up, in order to conserve energy, but this is useless because it only warms the skin. Similarly a pre-event massage is sometimes used as a substitute for a warm-up, to reduce the risk of injury and to enhance performance. However, the little evidence there is suggests that even as much as 30 minutes of skilled massage is mainly of psychological benefit, and that an active warm-up produces better results.

Secondary effects of cold

- Muscular: muscle *impairment*; muscle *injury*.
- *Shivering*: *incapacity* in very cold water.
- Cardiovascular: *angina* is more prevalent in the cold – if the respiratory tract is protected from the cold, exercise can continue without angina, even with moderately vigorous exercise in severe cold; rise in *blood pressure* – increases risk of *stroke*; *myocardial infarction* (heart attack), *heart failure*.
- Respiratory: exercise-induced *bronchospasm* (EIB) – asthma provoked by exercise; *rhinorrhoea* (streaming runny nose) on return to warm room, which increases susceptibility to upper respiratory infection; uncontrollable *hyperventilation* (overbreathing) in sudden severe cold – even if fully submerged – which can lead to loss of consciousness, and/or disablement by *tetany* (the overbreathing causes the CO_2 to fall to such a low level in the blood and brain that the person becomes *unconscious* and can also develop spasm-like epileptic *fits*).
- Nervous system: loss of manual *dexterity* and *sensitivity* (hand movement is clumsy and hands are less able to feel things); impairment of *co-ordination*; reduction in *alertness* and *visual acuity* (accuracy of what is seen); slowing of *reflexes*; increase in *mistakes*; *misinterpretation* of visual and auditory sensory input (sights and sounds); *hallucinations*
- Other: increased risk of '*bends*' (decompression sickness) in divers; *allergy* – possibly just a rash, for example, on legs after paddling in cold water, but may result in anaphylactic shock (allergic collapse as occurs in some people after eating peanuts); *snow blindness* – not due to cold but to excessive UV light reflecting off snow in bright sunlight. Dark glasses should be worn.

The active warm-up may provide real value because it fully activates the biochemical assembly line producing energy in the cells that are going to be used. The blood vessels to the active muscles will also have opened up, helping to spread the heat and ensuring optimal transport of food, oxygen and waste products.

Shivering

Shivering, which is often a feature of actual or incipient hypothermia, can result in reduced manual dexterity and can compromise skilled movements, and consequently increase danger in difficult situations. For example, the individual may not be able to keep strict control of his movements when shivering, and someone such as a skier may therefore have a reduced performance. Shivering will also produce a disastrous fall in performance in events in which steadiness is essential, for example, shooting and biathlon.

Exercise-Induced Bronchospasm (EIB/Asthma)

The incidence of EIB – symptoms include breathlessness and a 'tight' chest – is three times greater in cold-weather athletes compared with their warm-weather counterparts (*see also* Chapter 10). Experimentally, bronchospasm can be induced in all subjects with a combination of cold air and high respiratory minute ventilation (VE) (volumes of air breathed in and out in one minute). Different people have different sensitivities for the development of EIB.

EIB is triggered by cooling and/or drying the lining of the airways. Exercise increases ventilation. The temperature of the ambient air is normally lower than the core temperature and ambient air is much drier than air in the lungs, which is fully saturated with water vapour. EIB can be prevented by breathing saturated air at body temperature. EIB is related only to the respiratory minute VE and the cooling capacity of the air, and not to the exercise itself.

The risk of EIB/asthma is increased during a respiratory infection, and reduced by improved fitness and by breathing warm moist air. Asthma is no reason for not participating in exercise or sport.

NB: for Olympic or International level competition, athletes need a Therapeutic Use Exemption (TUE) certificate if they are using anti-asthma medication. To get the TUE certificate they must undergo an IOC-/WADA-recognized test to prove that they have reversible bronchospasm.

It has been shown that breathing air with a temperature down to –25°C does not lead to freezing of the lungs. Indeed, the ambient air temperature of –25°C is warmed to at least 25°C when it is 9cm inside the nasal passages (Webb, 1951), and in one ski race the air temperature was –28°C, with a wind-chill equivalent of –55°C, with no respiratory damage. However, with maximal exertion in very severe cold a bronchorrhoea may develop, and this is called 'freezing the lungs' by the Inuit of the Arctic. A similar problem can occur in horses and dogs in a similar maximum exercise/severe cold situation (Lloyd, 1986).

HIGH ALTITUDE AND COLD

An individual who can sustain high-intensity exercise may be able to generate sufficient heat to offset heat loss to the environment. At high altitude it may not be possible to carry out high levels of exercise because of the relative hypoxia, and exhaustion may limit the shivering response. Also, heat production is about 10 per cent lower while the individual sleeps, compared with being awake. Consequently the high-altitude climber has to rely more on clothing to protect himself from the cold.

COLD WATER

A number of circumstances may result in water immersion – the canoeist or sailor may fall into water, a cross-country skier may fall through ice, and a Channel swimmer may spend many hours swimming in the water. Open-water swims are carried out in the English Channel and in some lakes, where the late summer and early autumn temperatures are about 17–18°C. Triathletes also take part in open-water swims.

Each year in the UK over the last 25 years there have been between 400 and 1,000 deaths related to water immersion (Tipton et al., 1999). Many of these deaths have occurred despite the fact that a large number of the victims appeared to be competent swimmers. Possible causes for the deaths include drowning, hypothermia, uncontrollable hyperventilation, and impairment of swimming performance resulting from the adverse effects of cold.

Factors other than water temperature will influence the way in which core temperature responds to cold-water immersion. Air movement increases cooling and similarly, cold-water movement will promote heat loss. When the body is stationary in water, the water immediately surrounding the body is warmed and provides a protective layer. When this 'warm' layer of water is disturbed, it is replaced with cooler water and consequently there is a greater cooling effect. The layer can be disturbed by the flow of a river, or when a subject moves in the water by swimming or treading water.

Body size and body fat are relevant (*see* page 74), but the time of useful consciousness can be increased dramatically by wearing some form of body protection. Wearing ordinary clothing in water of 5°C may only allow an hour of useful consciousness whereas this time may be extended by 3 or 6 times by wearing a wet and dry suit respectively.

Sudden Immersion in Ice-Cold Water

If an individual is suddenly immersed in ice-cold water, the major aim is to survive the cold shock response and the associated uncontrollable hyperventilation (overbreathing), which may last for up to 3 minutes. This hyperventilation may increase the volume of air breathed in and out per minute to 100L/min with a respiratory rate of 60 breaths/min. (Normal resting values are 4–15L/min for the volume of air breathed in and out per minute and 8 to 15 breaths/min for respiratory rate.) Since breath hold time is reduced to around 10 seconds, a submerged individual may not be able to prevent a large influx of water into his mouth. During the uncontrollable hyperventilation there is also disruption to the normal coordinated linkage between swimming movements and breathing. Thus, the individual is likely to 'take in' water during this period if attempts to swim are made.

Tactics to survive the uncontrollable hyperventilation include a concerted effort to raise the head above the water line as soon as possible. There should be no attempt to swim for shore, or clamber to a river bank. It is better to cling to ice or a log and wait for the hyperventilation to subside.

The hyperventilation results in a reduction in CO_2 content in the body and possible reduction in blood flow to the brain resulting from the respiratory alkalosis (the blood becomes more alkaline because of the decrease in CO_2 body stores resulting from an increased volume of air breathed in and out per

Tactics to survive uncontrollable hyperventilation

- Make a concerted effort to raise the head above the water line as soon as possible.
- Do not attempt to swim for shore.
- If possible, cling to the side of ice or a log and wait for the hyperventilation to subside.

minute). This reduced blood flow to the brain may lead to confusion and poor decision-making, and, ultimately, to drowning. The hyperventilation also results in a feeling of breathlessness, and may promote a panic reaction.

The cold shock response appears to be most severe when the skin of the chest and back are exposed to cold water, and it will therefore be diminished by clothing on the trunk. A life jacket that keeps the head out of the water will lower the potential for water aspiration.

Very cold water in the ears or nose may case instantaneous cardiac and respiratory arrest. Cold in the ears may also lead to disorientation, which can cause submerged people such as capsized canoeists to swim downwards instead of towards the surface.

Figure 19 provides a flow chart of some possible consequences of immersion in cold water.

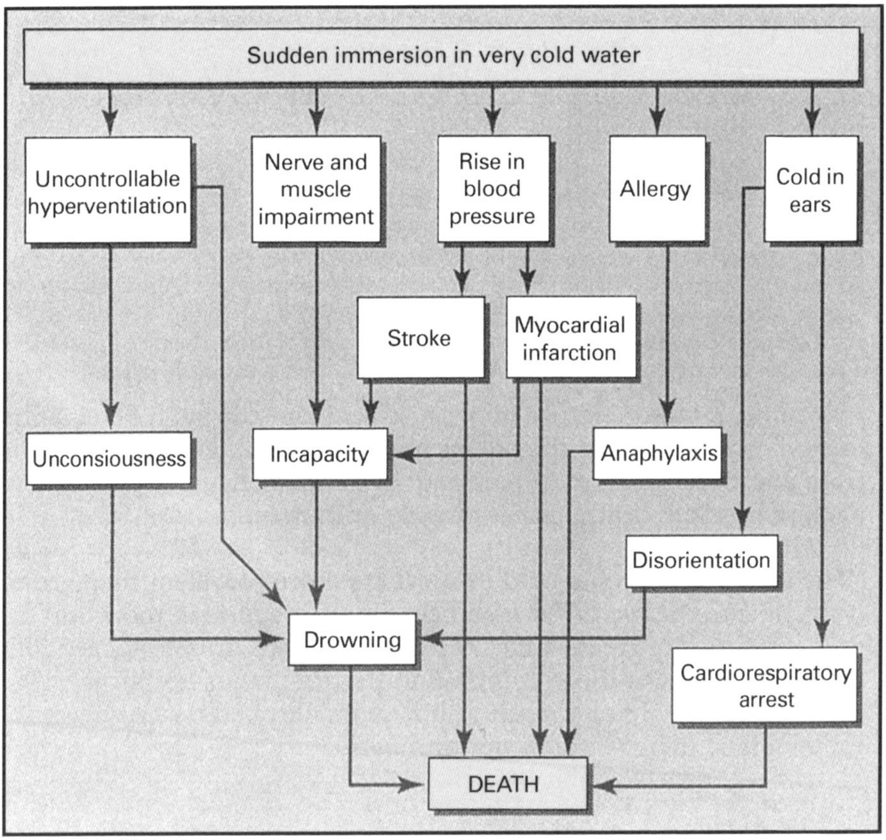

Fig 19. Flow chart of potential consequences of water immersion.

(Originally published in Lloyd, E. L. (1994), ABC of Sports Medicine: Temperature and Performance I: Cold, *British Medical Journal*, 309, 531–4. Reproduced with permission from the BMJ Publishing Group)

Long-Distance Swims

> . . . we found swim failure in cold water to be a progressive decrease in swimming efficiency in the absence of general hypothermia.
>
> (Tipton et al., 1999)

In the Tipton et al. (1999) study, ten swimmers were asked to complete a 90-minute swim in 25°C, 18°C and 10°C water on different days. All ten swimmers completed the swim in the 25°C trial but only eight and five swimmers completed the 18°C and 10°C swims respectively. At 10°C swimming efficiency was reduced, with a decrease in stroke length, an accelerated stroke rate and an increased swim angle. Limb cooling is deemed to be a major limiting factor in the ability to sustain swimming and to maintain good technique. During swimming, the arms are the main source of propulsion and therefore require a large blood supply, resulting in greater heat loss from this area.

Swimming speed and water temperature can influence the oxygen uptake (VO_2) of swimmers. Comparison of VO_2 in water temperatures of 18, 26 and 33°C showed that VO_2 was higher for 18°C than 26°C, which in turn was higher than 33°C at low swimming speeds, but at higher speeds VO_2 was similar for all three temperatures (Nadel et al., 1974).

The higher VO_2 at the lower water temperatures was attributed to shivering, suggesting that, in cold water, if the swimmer can swim fast, at a high intensity, he will be able to produce enough metabolic heat to offset heat loss from the cold water (Nadel et al., 1974). While this strategy may be more effective for 'fat' swimmers than for lean individuals, it is likely that it will result in a relentless decrease in core temperature for thin subjects. Pugh, a very thin exercise physiologist, compared his responses to lying and swimming in 16°C with those of a fatter English Channel swimmer. Pugh's core temperature fell by a little over 1.0°C in 80 minutes while he lay in the water, but he experienced a decrease of over 3.0°C in less than 80 minutes when he swam. The Channel swimmer showed about 1.0°C fall in core temperature while he lay in the water, and little change in core temperature during a 120-minute swim (Pugh and Edholm, 1955). It seems that a 'fat' swimmer is likely to cope better with cold water than a 'thin' swimmer.

Swimmers who aspire to carry out long-distance swims in water temperatures of 16–18°C are more likely to succeed if they have a protective layer of fat and are able to sustain swimming speeds great enough to minimize the cooling effect from the water. Exercise in water may not be adequate to prevent

Energy costs

- At slow swimming speeds in cold water, the energy cost is higher than in warmer water.
- The increased energy cost is attributed to shivering.
- Only when the speeds increase and heat production is higher (resulting in no shivering), do the VO_2 values converge.

(Based on the study by Nadel et al., 1974)

Cold-water strategies

- Lean subjects are particularly prone to cooling in cold water, so they should take special care when planning open-water swims.
- Exercisers who may fall into iced water should be familiar with the likely consequences and be aware of the appropriate actions to promote survival. For example, they should attempt to hold on to a floating object or ice until the cold shock response subsides.
- If stranded in a hole in the ice, the subject should lean on his arms; if unconsciousness ensues, it is possible that he will not sink in the water as his arms will be frozen to the ice.
- After the cold shock response has subsided, if land or a boat are close by (about 100 metres), it may be possible to swim to safety. Otherwise, lean individuals in particular should restrict limb movement.
- Before a long swim, the swimmer should consider increasing body fat, as long as he will not have to compete in running races too.
- Some form of protection should be worn in open-water swims. The arms should be covered and consideration should be given to wet or dry suits. Lowden (1984) recommends at least a short leg and vest top for outdoor swims. He also emphazies the need for a hat or cap that covers the ears.

a decrease in core temperature, and lean individuals are vulnerable. The arms should be 'protected' from cold water as they are a potent source of heat loss when an individual swims in cold water.

STRATEGIES TO COMBAT COLD

Avoiding Danger

Athletes must be aware of the circumstances that predispose individuals to the dangers associated with cold. There are a number of strategies that may be employed to deal with dangerous circumstances:

- Convective heat loss (wind) – avoid head winds in training, wear windproof clothing.
- Conductive heat loss – avoid contact with cool surfaces. Insulation from the ground (ice, snow, wet surfaces) – gloves should provide sufficient protection to avoid conductive heat loss while holding ski poles or rifle. Stay dry or make every effort to become dry to avoid heat loss.
- Radiative heat loss – reduce the skin surface area exposed to air.

Fig 20. A wet suit is desirable for open-water swims.
(Reproduced with permission from Charles Campbell, aka Charlie Campbell; *see also* www.munrorun2000.co.uk/gallery)

Exercising in Cold-Weather Conditions

There are a number of general strategies to follow when exercising in cold-weather conditions:

- Risk assessment and prevention strategies – before venturing into a cold environment, ask the following questions: What are the hazards – cold, wind, wet? Who is at risk – lean athletes, children?
- Evaluate the weather conditions (air temperature, wind chill, wind direction, rain and snow), and plan the direction of training runs accordingly. Where possible, choose a route that protects from a head wind but provides a tail wind in the other direction. If this is not possible, run into the wind on the way out and have the wind behind on the way back. The route and length of session should take into consideration the weather and underfoot conditions. Sessions should be shortened if fatigue may lead to vulnerability to the cold.
- If stationary before, during and after training or competition, avoid exposed windy situations and seek shelter from wind, rain or snow. Adjust clothing immediately when stopping.
- Avoid slippery surfaces. Slow down or stop if there is danger of falling on icy surfaces.
- In the dark or in dull days, wear bright reflective clothing. Do not wear white in snowy conditions, and avoid unlit locations. Run facing oncoming traffic. Be aware of potential traffic difficulties – read the road ahead and take avoiding action early. Do not be brave. Look for escape routes when traffic may threaten your safety.
- Stay dry if possible. Avoid a moisture build-up from sweating (through over-dressing), from rain or melting snow, or from contact with (for example, immersion in) cold water.
- Planning – have not only appropriate equipment for an emergency, but also a strategy to counter emergency situations.
- Be aware of the possibility of poor decision-making – judgement can be impaired in cold conditions. Train with a friend, take a mobile phone, or leave behind a note of the planned route.
- All exercisers should be aware of warning signs, both in themselves and in each other. Signs of hypothermia include shivering, unusual behaviour, unaccustomed slow responses, slurred speech.
- Ask each other about toes and fingers, in case of frostnip.

Clothing

- The exerciser should be aware of the characteristics of clothing, and find a clothing ensemble and strategy that suits them and the prevailing conditions. The aim is to remain comfortable in a range of cold environments.
- It is usual for the athlete to feel quite cold in the first 5–10 minutes at the beginning of the exercise session. If he does not feel cold at the start of exercise, he is probably overdressed. He should not expect to feel comfortable until he has been operating for a few minutes at training pace.
- Clothing should have layers, and each layer and materials should have a specific role: the inner layer should be able to wick fluid from the skin; the

Fig 21. Runners setting off in cold conditions.

Once they have warmed up, as a result of heat generation from the working muscles, they may wish to take off their hats, and remove or adjust some clothing. (Reproduced with permission from Charles Campbell, aka Charlie Campbell; *see also* www.munrorun2000.co.uk/gallery)

insulative layer or layers should provide protection from cold; an outer layer should be wind- and waterproof.
- Clothing should be adjustable to avoid overheating while exercising. Ventilation areas should be closed during rests or while running or skiing into a head wind.
- Expose areas that promote heat loss (for example, the head) when exercising to avoid an unnecessary build-up of heat, and cover potent areas of heat loss when resting.
- Overdressing results in unnecessary sweat production and unnecessary wetting of clothes.
- The materials should suit their particular role – for example, polypropylene wicks away fluid, while cotton and down lose a great deal of their insulation when wet.
- Special attention should be given to the extremities.
- It is highly desirable to avoid wet feet.
- The head and neck are particularly vulnerable to heat loss as the blood vessels in these areas do not constrict very well in cold.
- Shoes should not restrict blood flow to the toes. Thick soles and insoles help restrict heat loss.

Fatigue

It is vital to be aware of the dangers of fatigue. In a cold environment the tired athlete may not realize that his slow pace and low heat generation can lead to hypothermia, impaired judgement and an inability to carry out simple tasks such as adjusting or changing clothes.

Aerobic Fitness

Individuals with a high level of aerobic fitness are able to produce more metabolic heat than unfit individuals; this allows the fit individual to sustain high levels of heat production for long periods.

Food, Water and Alcohol

- Athletes should carry carbohydrates on long training runs in the cold.
- Dehydration increases the chances of frostbite and hypothermia.
- A 'cold' athlete should not be given alcohol, as alcohol promotes heat loss by dilating blood vessels.

CHAPTER 6

Altitude

Some athletic events are held at altitude – the 1968 Mexico Olympic games took place at 2,300 metres; rugby in South Africa may be played at 1,800 metres; cross-country skiing races sometimes take place at 2,000 metres. Many athletes believe that living at altitude and training there, or training near sea level, but living at altitude, has the potential to enhance sea-level performance.

ATHLETIC PERFORMANCE AT ALTITUDE

Performance in a range of anaerobic events is likely to be enhanced at altitude because air density is less. Anaerobic events do not rely on the O_2 transport system. Analysis of performances in the Mexico Olympics highlights that anaerobic performance was improved relative to the world records of the time; several sprint world records – in the men's 100m, 200m, 400m, 800m and 400m hurdles – were set at the 1968 games. However, times in the aerobic events in the Mexico Olympics were well below the world records of that time; the times for the 3k steeplechase, 10k and Marathon were slower by 4 per cent, 7 per cent and 8.5 per cent, respectively.

It is also worth noting that at altitude (for example, at around 2,000m), the flight characteristics of a football or rugby ball will be affected. The players may also be able to kick a ball further than at sea level.

Performance at altitude

Sprint performance may be enhanced at altitude, and endurance performance is worse at altitude.

BODY RESPONSES AT ALTITUDE

For a more detailed explanation of the responses of the body to altitude, using physiological terms, *see* Appendix II.

The barometric pressure decreases at altitude and therefore there is a lowering of oxygen pressure. The lowering of oxygen pressure results in a reduced oxygen in the arterial blood, particularly above 3,000m. At rest, at around 1,300m, the oxygen content in the arterial blood will be similar to that at sea level. Thus, while the oxygen pressure falls with increasing altitude, the arterial blood 'holds on' to the oxygen fairly well at low altitudes. The fall in oxygen pressure is a stimulus for the individual to increase the amount of air that he breathes in and out. This is generally advantageous, as it increases the arterial oxygen content. Some individuals have a more marked increase in volume of air moved in and out of their lungs than others. The increase in the volume of air breathed in and out per minute makes the individual more breathless during exercise at altitude.

Immediately on ascent to altitude, the body attempts to adjust to the decreased oxygen in the arterial blood by increasing the rate of flow of blood pumped by the heart. At submaximal levels, this is reflected in an increased heart rate. This compensation allows the individual to carry out submaximal exercise that can be achieved at sea level, but at maximum or near to maximum, it is not possible to enhance the blood flow from the heart sufficiently, and a decrease in maximal oxygen uptake (VO_2max) is apparent at altitude.

The increase in blood flow from the heart is considered to be an expensive way of countering the reduced arterial oxygen content. During the first week at altitude, the increased blood flow from the heart decreases and the skeletal muscles increase the amount of oxygen that is extracted from the blood.

Haemoglobin carries most of the oxygen in the blood and the haemoglobin concentration of the blood is increased at altitude. Initially, the increase is a result of a reduction in the watery part of the blood, the plasma. Over time, the low oxygen is a stimulus for more red cells to be produced, resulting in an increase in haemoglobin and an enhanced ability of the blood to carry oxygen.

EXERCISE AT ALTITUDE

The oxygen saturation in the arterial blood (SaO_2) at 4,500m at rest is around 80 per cent. Light exercise at 4,500m has been shown to lower SaO_2 to 70 per cent. The resultant reduced O_2-carrying capacity of the arterial blood contributes to a decrease in VO_2max in exercise at altitude. Noakes (2001) expounds the theory that endurance performance is reduced at altitude because there is limited delivery of oxygen to the heart. Consequently, the brain limits the amount of skeletal muscle that can be recruited during maximum exercise resulting in a reduced performance.

There is great individual variability in the effect of altitude on the VO_2max. The altitude required for a reduction in VO_2max has been shown by Gore et al. (1997) to be as low as 580m in male and female athletes. At the end of the VO_2max test, the male athletes had an SaO_2 of 92.0 per cent at sea level and 90.1 per cent at 580m, and at 580m, had a 6 per cent reduction in VO_2max compared with sea level. At 580m, the female athletes had a 4 per cent reduction in VO_2max, and a decrease in SaO_2 from 92 per cent to 90 per cent compared with sea level.

It would appear that those athletes who have a 'low' SaO_2 at the end of a maximal test at sea level are more likely to have a reduced VO_2max at low altitudes, compared with those who have a high per cent SaO_2 at sea level. Thus, some athletes are likely to experience a decrement in VO_2max at very

Exercise at altitude

- At altitude, VO_2max is reduced and subjects exercise submaximally at a higher relative intensity compared with sea level.
- The higher volume of air breathed in and out per minute results in a greater perception of breathlessness.
- There is a marked individual response to aerobic exercise at altitude.

low altitudes and are more likely to be susceptible to decreases in performance, compared with others.

At 2,300m a decrease of around 15 per cent in VO_2max is to be expected, whereas at 4,000m the fall in VO_2max compared with sea level is about 27 per cent. Thus, if an athlete exercises at 50 per cent of his VO_2max at sea level, the same absolute intensity at 4,000m would be around 70 per cent of the VO_2max, meaning that he would be exercising at higher relative intensity at altitude. At altitude the VO_2max is reduced, and subjects exercise submaximally at a higher relative intensity compared with sea level. Thus, it is not surprising that the athlete will feel that the same speed at altitude will be more stressful than at sea level. Again, it is emphasised that there is a marked individual response to altitude. The 'fittest' athletes may experience the greatest decrement in performance at altitude.

In three separate studies, decreases in VO_2max of 14, 16 and 17 per cent compared with sea-level values have been reported on Day 1 at 2,300m. In the first two studies, after 16 and 45 days at 2,300m respectively, the VO_2max decrement was 11 per cent, while in the third study there was a 15 per cent decrease in sea-level VO_2 max after 20 days at altitude. These data suggest that an athlete at 2,300m can expect about 15 per cent reduction in sea-level VO_2max but that there may be large individual differences. Similarly, improvements in VO_2max after residence and training at altitude are likely to vary considerably. These improvements in some studies have been meaningful, but the VO_2max is still well below sea-level values.

The studies that have assessed the effects of altitude residence and training indicate that endurance performance is reduced at altitude compared with sea level but that, over time at altitude, endurance performance may improve but does not return to sea-level values. Examples of 1-, 2- and 3-mile performance changes at 2,300m are shown in Table 8.

GUIDELINES FOR TRAINING AT ALTITUDE

Residence and training at altitude are likely to enhance endurance performance at altitude. The fact that some races are won by very small margins shows that training and staying at altitude is worthwhile for the athlete who will compete at altitude.

Guidelines for those athletes who wish to compete at altitude suggest that the higher the altitude of the competition venue, the longer the athlete should

Table 8 **Performance at altitude before and after acclimatization**

Study	*Distance*	*VO_2 max as % of SL*	*per cent of SL*
Adams et al. (1975)	2 miles	107 (3)	105 (18)
Pugh (1967)	1 mile	104 (Wk 1)	102 (Wk 4)
	3 miles	108 (4)	106 (29)
Faulkener et al. (1967)	2 miles	107 (14)	106 (32)
Daniels & Oldridge (1970)	1 mile	106 (Wk 1)	105 (Wks 3 & 4)
	3 miles	110 (Wk 1)	109 (Wks 3 & 4)

(SL = performance at sea level. The figures in brackets represent the days or weeks at altitude when the test was carried out.)

(Adapted from Adams et al. (1975), Effects of equivalent sea-level altitude training on VO_2max and running performance, *Journal of Applied Physiology*, 39, 262–6. Used with permission from the American Physiological Society.)

remain at altitude before competition. Smith and Sharkey (1984) suggest that athletes should spend 14 days at 2,000m, and 28 days at 2,500m before competition at these heights. Recommendations from other sources are similar, with 3 weeks' acclimatization suggested for competition at altitudes of 2,000 to 2,500m. Aerobic training speeds will be reduced and it may not be possible to complete the same number of intervals that were performed at sea level. Consideration should also be given to longer recovery periods between intervals in an attempt to maintain quality.

While athletes stay at altitude to acclimatize for competition, it may be prudent to train low on some occasions to avoid a possible detraining effect. Although there appears to be no experimental data to support training at sea level as part of the preparation for competition at altitude, this strategy may prove to be effective (*see* below). It has been suggested that athletes who are unable to find the time to acclimatize for competition at altitude should arrive at altitude on the day of the event. This advice applies to athletes who are taking part in predominately anaerobic events. Before travelling to altitude, an athlete may be able to help the acclimatization process by sleeping in an altitude house or taking part in 'altitude' training sessions in an environmental chamber.

Athletes should consume iron-rich foods before and during altitude exposure to ensure that there is no inhibition of red-cell production. Iron stores should be checked before travel to altitude. Residence at altitude may suppress appetite and, consequently, carbohydrate intake may be reduced. The same running speed at altitude will be at a higher relative intensity than at sea level and consequently is likely to increase glycogen usage and lead to reduced endurance. Thus, emphasis should be given to carbohydrate intake at altitude.

Dehydration can reduce endurance performance. The increased minute ventilation (VE) at altitude results in a high water loss as the air is dry at altitude,

Guidelines for athletes planning to compete at altitude

- Highly trained athletes are likely to improve endurance performance at altitude with residence and training at altitude.
- The higher the altitude of the competition venue, the longer the athlete should remain at altitude before competing.
- Athletes should live at altitudes above 2,100m.
- Athletes who wish to perform at altitude should train at altitude and at sea level.
- It may be prudent to train low on some occasions to avoid a possible detraining effect.
- Athletes taking part in predominately anaerobic events who are unable to find the time to acclimatize for competition at altitude should arrive at altitude on the day of the event.
- The worst length of stay before competition is 3–6 days (Noakes 2001).
- High-intensity training should be avoided in the first 3 days. It may not be possible to complete the same number of intervals as at sea level.
- Altitude houses or environmental chambers may be used before travel to help the acclimatization process.
- Athletes should consume iron-rich foods before and during altitude exposure.
- At altitude, games players need time to adapt to the new flight characteristics of the ball, and the change of pace of the game.
- Dehydration may reduce endurance performance; the air is dry at altitude and the athlete must add moisture to this dry air.
- At altitude, athletes will be susceptible to sunlight and should protect their skin and eyes.

and the athlete must add moisture to this dry air. In the clean 'thin' air, athletes will be susceptible to sunlight, and should protect their skin and eyes.

PRECAUTIONS FOR LIVING AT ALTITUDE

Athletes must drink 'extra' fluid at altitude because the dry air means sweating may be unnoticed. There is also increased fluid loss, as a result of breathing an increased volume of air in and out per minute in a dry atmosphere.

During the first few nights, athletes may have symptoms of Acute Mountain Sickness (AMS) (*see* below), and training should be adjusted until the symptoms disappear. The use of Diamox (acetazolamide), as for the prevention of AMS, may be of value. If the diuretic (increased production of urine) effect is considered undesirable, dexamethasone could be substituted.

Athletes of African origin or ancestry must be checked for sickle-cell disease since a decreased pressure of oxygen (PO_2) triggers sickling, with many pathological effects including, occasionally, splenic rupture. In the USA 13 per cent of people of African origin show a sickling trait (*see* Chapter 10, and Glossary).

Haemoglobin and serum iron levels should be checked regularly and treatment given if required.

The effects of increased ultraviolet (UV) light at altitude must be considered and countered. Sunscreens and dark glasses must be used.

ALTITUDE TRAINING AND SEA-LEVEL PERFORMANCE

Potential Benefits

Many elite endurance athletes consider that altitude training is an essential component of their training programme. Paula Radcliffe, the world's fastest woman marathoner in 2005, is one athlete who spends a considerable time living and training at altitude. Many world-class endurance athletes come from East Africa; they were born at around 2,500m and ran several miles to and from school each day. This suggests that altitude training has the potential to enhance sea-level performance, but it is necessary to examine the scientific literature to determine whether living and training at altitude is beneficial for sea-level endurance performance.

One adaptation of living at altitude is an increase in haemoglobin, resulting in an increase in arterial oxygen-carrying capacity. Reinfusion of blood (also known as blood doping, and a banned procedure in sport) has been shown to increase the VO_2max at sea level. It seems plausible that residence at altitude, resulting in an increase in haemoglobin, could improve endurance performance at sea level. However, some studies do not support this reason for an enhanced sea-level performance.

Considerations

The threshold altitude for a decrease in the VO_2max has been shown to vary between studies. The VO_2max of elite athletes has been reduced by 5 per cent at 580m, although some only found a fall in VO_2max at 1,200m. Above the threshold for a decrease, the greater the altitude, the greater the decrease in VO_2max. After initial exposure to altitude, VO_2max is reduced by about 15 per cent at 2,300m, and about 27 per cent at 4,300m. Differences in the reductions in the VO_2max are due to individual variation in response to altitude. The decrease in VO_2max is usually less after a stay at 2,300m. However, some studies have shown no improvement in the VO_2max after some days spent at around 4,000m. Distance running performance is decreased at altitude – 1- and 2-mile times have been found to be 4–7 per cent slower at 2,300m compared with sea level, with improvements of 1–2 per cent after 2–4 weeks at altitude. In addition, it may not be possible to complete the same number of intervals at altitude as at sea level. Athletes may not be able to train at normal sea-level speeds, and detraining could result. Attempting to avoid altitude illness, by staying at a relatively low altitude, may reduce the stimulus for red-cell production, while too long a stay at altitude may result in boredom and detraining.

Some athletes respond more favourably to altitude exposure than others. Possible reasons could include the range of iron stores in the body. Iron is an essential component for the development of red cells. Thus, athletes with low iron may not benefit from altitude exposure. Differences in the respiratory response to altitude may also be a contributory factor to differences in haemoglobin increases. A very marked increase in the volume of air breathed in and out in one minute at altitude will defend the decrease in SaO_2 and result in a blunted erythropoietin release.

Athletes should also be aware that the stress associated with altitude training may result in overtraining problems. There is also some evidence that the immune system can be suppressed, resulting in an increase in upper-respiratory tract and gastro-intestinal infections.

The Research

It may seem easy to determine whether altitude training is beneficial for sea-level performance simply by examining the scientific literature. However, the research should be interpreted with caution. Many studies have had limitations. They may have had no control groups – only the athletes who went to altitude were studied – and these athletes may have improved not because of the altitude, but because they were involved in increased training, and/or they responded well to the attention they received from experimenters. Several studies have had a low number of subjects, so the findings are not really representative; the results from a single subject may have had a great effect on the scores. Other studies have used recreational athletes, and it is not possible to conclude that the same findings will apply to elite athletes.

Well-designed studies are characterized by random assignment to an intervention group (altitude group) and a control group. The athletes should be highly trained so that the influence of additional training is unlikely to have a bearing on the results. Indeed, some studies have included a run-in period, whereby all of the athletes underwent the same training regime for several weeks before they were allocated to a study group in an attempt to avoid an added training stimulus influencing the results.

It is difficult to summarize research into altitude training, as there have been many studies with conflicting results, and altitude and time at altitude have varied greatly. However, below are some examples of altitude studies.

After 10 days at 2,300m, five moderately trained subjects (Balke et al., 1965) improved their VO_2max and 1-mile performance time. The fact that the mile times were over 5 minutes indicates that the subjects had limited ability and suggests that the improvement may be attributed to additional training and not the benefits of altitude training. Daniels and Oldridge (1970) exposed six elite runners to 2,300m for 14 days on two occasions with 5 days at sea level between exposures. After the second 14-day period at 2,300m and a further 5 days at sea level, the athletes spent another 7 days at altitude (on two occasions with an 11-day period at sea level). The athletes produced 14 personal bests and one (Jim Ryun) broke the world record for the mile.

As the above studies did not have a control group, it is difficult to attribute any benefit to altitude training. However, in the Daniels and Oldridge (1970) study, following their periods at altitude, elite athletes improved their VO_2max at sea level, and ran personal bests at sea level.

Some controlled studies have demonstrated improved sea-level performance but others have failed to demonstrate a benefit at sea level from altitude training. For example, the study by Adams et al. (1975) used a control group to evaluate the effect of altitude training at 2,300m. The authors concluded that there was no meaningful benefit from training at altitude. There is limited evidence from controlled studies to support altitude residence and training at altitude for enhanced sea level performance.

Altitudes under 2,100m may provide insufficient stimulus to increase haemoglobin mass. Even altitudes above 2,100m may not promote an increase in haemoglobin. Gore et al. (1998) studied world-class athletes and found that exposure to an altitude of 2,700m did not lead to an increase in haemoglobin or the VO_2max at sea level. Ri-Li Ge et al. (2002) concluded that, for the majority of subjects, the threshold of between 2,100 and 2,500m had to be reached before an increase in erythropoietin was triggered. However, some subjects increased erythropoietin at 1,700m, whereas others experienced no increase at 2,800m. The authors speculate that differences in the SaO_2 among individuals are a contributory factor to the individual erythropoietin response. While the relationship between SaO_2 and erythropoietin was not particularly strong, a lower SaO_2 was associated with a higher increase in erythropoietin. Differences in altitude and subject characteristics are possible explanations for the lack of consensus about an appropriate altitude for an increase in haemoglobin mass.

In their review of altitude training, Rusko et al. (2004) refer to the studies in the area when they recommend at least 12 hours per day at an altitude of 2,100m to 2,500m for at least 3 weeks to increase haemoglobin mass. However, 3–4 weeks at the above altitude range may not apply to trained endurance athletes who have a greater haemoglobin mass and blood volume than untrained subjects. The lack of consensus about the altitude and length of time at altitude that will promote haemoglobin mass increases is probably due to the fact that individual responses to increases in haemoglobin mass are markedly different.

Roberts et al. (2003) state that studies that have used the live high, train low (LHTL) strategy (*see* page 106) have generally reported performance enhancement of between 1 per cent and 4 per cent in events between 400m and 5k. These improvements are clearly of benefit to the athlete and there are a number of possible reasons for the enhanced performance. In their review, Hahn and Gore (2001) concluded that the research findings indicate that altitude training is unlikely to enhance sea-level endurance performance as a

Increasing haemoglobin mass

- Have an iron status check and take iron supplementation if necessary.
- Live between 2,100 and 2,500m for at least 12 hours per day for at least 3 weeks.
- Individual variability suggests that the above recommendations for altitude and time at altitude may not be appropriate for all.

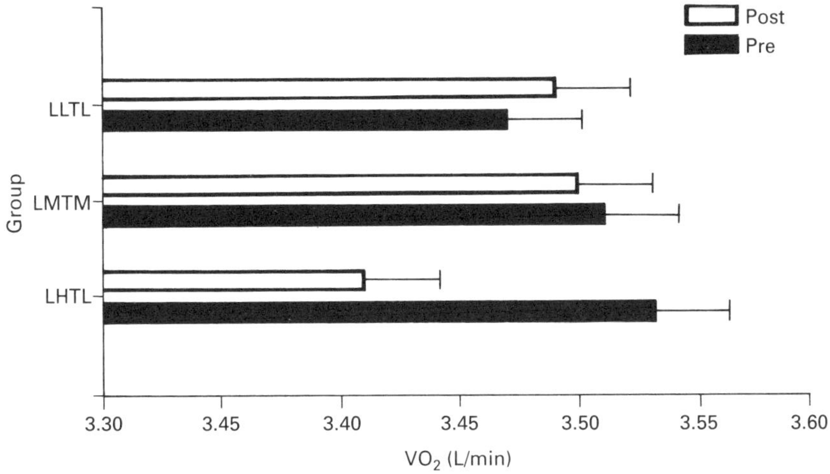

Fig 22. Oxygen uptake (VO_2) at a given running speed for three groups who carried out Live High, Train Low (LHTL), Live Moderate, Train Moderate (LMTM), and Live Low, Train Low (LLTL).

The post-test VO_2 is lower for the LHTL. (From Fig 1 in Saunders et al. (2004), Improved running economy in elite runners after 20 days of simulated moderate-altitude exposure, *Journal of Applied Physiology*, 96, 931–7. Used with permission from the American Physiological Society)

result of an increase in VO_2max in elite athletes, while Levine and Stray Gunderson (2001) attribute the enhanced sea-level performance to a VO_2max increase.

Hahn and Gore (2001) suggest that altitude training may improve sea-level performance by improving the mechanical efficiency of the athletes. For example, they have shown that their world-class cyclists use less oxygen at a given speed after altitude training. In addition, they found that the muscle buffering capability (ability to counter increases in acidity) was enhanced – in other words, the skeletal muscles were better able to counter an increase in acidity after altitude training.

Saunders et al. (2004) found that running economy was improved after a period of living high and training low (Figure 22). The reduced oxygen cost of running at a given speed is likely to enhance endurance performance, especially in long-distance events. There appears to be no study that has demonstrated a detrimental effect on exercise economy. Some have shown an improvement in cycling or running economy, and in some there has been no change. A recent study in swimmers found that 13 days training at 1,850m produced an increase in red cells that did not occur when the same swimmers trained at 1,200m for 13 days. However there was a greater improvement in performance after the period at 1,200m than at 1,850m (Roels et al., 2006). This would suggest that the improvement at 1,200m was not related to changes in blood components.

Strategies for Altitude Training Exposure

There are a number of strategies that have been adopted to improve sea-level endurance performance by training at altitude. The traditional method of altitude training is to live and train on a mountain. This method has been criticized as having limitations, including a decrease in training speeds and the possibility of a detraining effect. Some athletes may experience altitude illness. In an attempt to overcome some of these problems, some athletes have lived on the mountain but trained in the valley. Thus, they are able to gain the benefit of living at altitude – an increased haemoglobin concentration – but are able to train at their normal speeds.

The expense and inconvenience of travelling to mountains have prompted some groups to build altitude houses. In these houses the barometric pressure is decreased (simulating altitude), or the nitrogen (N_2) is increased and consequently per cent of O_2 in the air is reduced to produce a similar effect. The athletes live in the houses to gain the 'altitude' effect but are able to train at sea level outside without reducing training intensity.

Many athletes live and train at altitude to increase haemoglobin concentration and red cell mass. This increases the O_2-carrying capacity of the arterial blood, and enhances the ability to perform aerobic activities. Thus, those athletes who have an improved aerobic capability resulting from altitude exposure will be able to perform better (faster) at altitude, and at sea level. Consequently, the training stimulus will be increased. While increased haemoglobin mass is an important variable, other factors can also influence endurance performance, including speed at lactate threshold and running economy. Some factors may be adversely affected by training at altitude. During maximal exercise at altitude the electrical activity to the muscles and the maximum cardiac output are reduced, suggesting that inhibition by the central nervous system limits the utilization of the skeletal muscles. Thus, the power output of the skeletal muscles is compromised at altitude, leading to a reduced training stimulus. Athletes are generally unable to maintain sea-level training speeds. This may result in a decreased neuromuscular training stimulus, and they may therefore detrain at altitude.

The possibility of a detraining effect as a result of training at altitude has prompted some studies in which athletes live at altitude to gain the altitude benefits relating to enhanced red-cell formation, but train in the valley to maintain training speeds. Levine and Stray Gunderson and others used stay at altitude or altitude houses to show that 'Live high, train low', or LHTL, has resulted in endurance performance improvement that is greater than when living high and training high. Levine and Stray Gunderson attributed the enhanced 5k sea-level running performance (after LHTL) to an increase in erythropoietin, which stimulated an increase in red cells and an improved VO_2max, which was directly related to the increase in red cells.

However, the LHTL strategy is only effective for responders, and Chapman et al. (1998) highlighted some factors relating to such individuals. Responders were defined as those athletes who enhanced their 5k running time by more than the group mean; those who showed a reduced performance were categorized as non-responders. Responders showed a greater increase in erythropoietin with acute exposure to 2500m altitude and had a sustained elevation in erythropoietin after 2 weeks; in non-responders, erythropoietin decreased after 2 weeks at altitude. The increase in erythropoietin in the responders resulted in an enhanced haemoglobin mass when compared with the non-responders. The responders demonstrated an increase in VO_2max; there was no change in the non-responders. The responders had less of a reduction in training speeds and oxygen uptake at low and moderate altitudes than the non-responders.

Timing

There is a lack of scientific evidence on how long the possible benefits of altitude training/exposure may last, or indeed when they may begin after descent from altitude. The aim must be to arrive at competition with the increased oxygen-carrying capacity still present, but having allowed sufficient time at sea level to reverse the negative effects of altitude, namely alterations in buffers, and fluid loss. One athletics coach recommends 15–18 days after time at altitude for optimal sea-level performance, but no studies are cited to support this statement. The life of a red cell is 120 days, but on return to sea level there is an immediate decrease in red-cell production. However, studies have shown haemoglobin levels to be higher than pre-altitude exposure 10 to 12 days after stay at altitude.

Effects on Anaerobic Performance

A limited number of studies have investigated the effect of altitude training/exposure on anaerobic performance at sea level. Although the results of these studies are not consistent, there is a suggestion that anaerobic performance (maximum effort of about 300 seconds) can be improved with altitude training/exposure. Mizuno et al. (1990) showed that a 2-week stay at 2,100m, with training at 2,700m, increased short-term sea-level treadmill time (treadmill time of around 300 seconds), with no change in VO_2max. The authors concluded that altitude training can enhance short-term exercise performance, and speculate that the improvement may be attributed to an improved muscle buffer capacity.

The 'live high, train low' (LHTL) strategy has been shown to be effective in enhancing anaerobic performance. In one study (Numella and Rusko, 2000), elite athletes spent 10 days (16.5 hours per day) in an altitude house at an altitude equivalent to around 2,200m. The athletes trained at sea level over the 10-day period. The altitude house subjects improved their 400m times, while the control group did not. The authors speculated that the improvements in 400m performance were as a result of changes in lactate metabolism and acid-base balance.

An Australian study (Roberts et al., 2003) compared the responses of well-trained cyclists to LHTL (8–10 hours per night in an altitude house at a simulated altitude of 2,650m) of 5, 10 or 15 days and a sea-level control group. All subjects trained at 610m. There was no difference in the improvements in maximal mean power output in 4 minutes, VO_2max, and maximum accumulated oxygen deficit (an assumed measure of anaerobic capacity) between the altitude groups, but all of the altitude groups had improvements in these three variables compared with the sea-level control group.

Altitude summary

- Anaerobic performance may be enhanced at altitude where wind resistance is a factor.
- Anaerobic performance may be enhanced at sea level after exposure to altitude, altitude training and LHTL.
- Endurance performance is reduced at altitude.
- The reduction in VO_2max at altitude has been shown to occur as low as 580m.
- Residence and training at moderate high altitude normally results in an improved aerobic performance at altitude.
- Residence at altitude may result in sea-level endurance performance improvement depending on the strategy employed.
- The mechanism for the increase in endurance performance after living and training at altitude, or LHTL, may be attributed to a number of factors, including an increase in haemoglobin mass, an enhanced muscle buffering capacity and an improved exercise economy.

Guidelines for Altitude Exposure

Athletes, especially females, should have their iron stores checked before altitude exposure. If iron stores are found to be low, steps should be taken to increase them before going to altitude. Levine and Stray Gunderson (2001) have reported that 20 per cent of their male and 60 per cent of their female endurance runners had reduced bone-marrow iron stores. Iron-deficient athletes will not respond well at altitude with no increase in haemoglobin mass or VO_2max.

The response to altitude is individualistic, suggesting that experimentation with altitude training is worthwhile. However, although the difference in responses makes general guidelines difficult, recent research does indicate that athletes should adopt the live high, train low (LHTL) strategy.

There may be more than one mechanism that enhances sea-level performance. Increased haemoglobin and improved exercise economy and an increase in skeletal muscle enzymes are all possible. While debate continues on the possible mechanism for improved sea-level performance after altitude exposure, there is some agreement that the desirable altitude for a beneficial effect lies between 2,000m and 3,100m. This may be achieved using altitude houses, or residence at altitude. Saunders (Figure 22) showed running economy improvements after 20 days within this altitude range, and reported that athletes ran personal and seasonal bests for events between 1.5 and 10k. Gore et al. (1998) showed no increase in haemoglobin after 31 days at 2,700m, but cyclists who lived permanently at 2,600m had higher haemoglobin levels compared with those who lived at sea level. Rusko et al. (2004) recommend at least 12 hours per day at an altitude between 2,100m and 2,500m for at least 3 weeks to increase haemoglobin mass.

The fact that some studies have failed to demonstrate a statistically significant difference should not in itself deter athletes from attempting altitude training. Even in the so-called negative studies, it has often been the case that some athletes have shown sea-level performance improvement. Remember: races are often won by mere hundredths of a second.

ALTITUDE ILLNESS

Lowlanders who venture to altitude may experience altitude-related illness. The main illnesses are acute mountain sickness (AMS), high-altitude pulmonary oedema (HAPE) and high-altitude cerebral oedema (HACE). The rate of ascent, the height attained and individual characteristics will all have a bearing on the incidence of such illnesses. Normally the athlete will have a very fast ascent as he flies to a given altitude. Occasionally, there may be a staged ascent as in the Everest marathon, where the athletes walk progressively higher before they begin the marathon at about 5,000m.

Medication

It may be possible to avoid altitude illness by adopting a gradual ascent but consideration should also be given to taking medication, particularly for those who have been shown to be predisposed to altitude illness (*see* below).

Acute Mountain Sickness (AMS)

Incidence

While AMS is considered to be more likely at altitudes greater than 2,500m, it can occur between 1,500 and 2,500m (Basnyat and Murdoch, 2003). The AMS rate is generally higher with increasing altitude, and is greater the faster the rate of ascent to altitude. The incidence of AMS has been shown to be 49 per cent at just over 4,000m, 34 per cent at 3,500m, and 13 per cent at around 3,000m. While the height attained and AMS incidence relationship tends to hold true, there may be some inconsistencies. A possible explanation for these discrepancies may be due to differences in scoring systems and the timing of the assessment of the AMS. AMS symptoms often impose limitations on a subject so that he does not feel like climbing higher.

Some individuals succumb to AMS at moderate altitudes despite the fact that the oxygen saturation in the arterial blood (SaO_2) is over 90 per cent. The underlying mechanisms relating to AMS are unknown. Individuals whose response to altitude is a less pronounced increase in the volume of air breathed in and out per minute may be more prone to AMS, because the hypoventilation (reduced ventilation) results in an enhanced cerebral (brain) blood flow. Characteristics associated with AMS include brain swelling and fluid retention. Not only is AMS associated with fluid retention, there is also a redistribution of fluid with the body. Those who have AMS have less urine output compared with non-AMS subjects, who have a slightly increased urine output when they ascend to altitude. Another possible cause of AMS is a decrease in brain oxygen saturation. (Administering oxygen does not immediately reverse symptoms, however.)

Symptoms and Measurement

AMS results from exposure to altitude but the symptoms are not apparent immediately on arrival at altitude. There is a time delay of 6–12 hours after exposure, and symptoms are normally worst on Days 2 and 3, and usually disappear after 4 to 5 days. The symptoms do not present themselves at that altitude again on the same trip.

Altitude scientists at a meeting in Lake Louise in Canada devised a scale to measure AMS. Subjects score the symptoms from 0 to 3 (*see* below). Headache is the primary symptom, but they also identified other symptoms:

- blunted appetite;
- feeling nauseous, vomiting;
- fatigue, weakness, lethargy;
- dizziness/light-headedness;
- sleep disturbance.

Interpretation of the scale varies slightly with some scientists using a score of 3 and above with a headache as signifying AMS, whereas others have used a score of 4 and above with a headache. Headache is considered to be the primary symptom of AMS.

It is important to remember that similar signs and symptoms may be due to other causes or conditions (these are not in order of likelihood):

- carbon monoxide poisoning;
- dehydration;
- diabetes;
- exhaustion;
- hangover;
- migraine;
- viral illness;
- brain tumour.

Avoidance and Treatment

The faster the rate of ascent, and the higher the altitude, the more likely it is that AMS will result. There is significant individual variation and some trekkers can ascend around 500 to 600m per day without succumbing to AMS. However, it has been shown that AMS incidence is decreased by taking a slow approach on ascent altitude, so immediate ascent to high altitude should be avoided.

Above 3,000m, subjects should sleep no higher than 300m above the previous sleeping altitude, and every few days, two nights should be spent at the same altitude. For the 'sea-level' athlete who proposes to live at 2,500m, it may be worthwhile spending one night at 1,800m before ascending to 2,500m.

Once a subject has presented with AMS symptoms, he should not climb any higher, and descent should be considered if no improvement takes place.

Diamox (Acetazolamide) is the preferred drug to combat AMS, but there is ongoing debate as to the optimal dose (Basnyat and Murdoch, 2003). Diamox counters AMS by increasing urine production and by making the body more acid, which stimulates the body to breathe more. The resultant increase in volume of air breathed in and out per minute increases the SaO_2 and decreases pressure of CO_2 (PCO_2). It is recommended that Diamox be taken 24 to 48 hours before ascent to altitude, and at altitude. Side effects include tingling in hands and feet and an increased urine output.

Diamox 250mg twice or three times per day and 4mg of Dexamethasone every 6 hours reduce AMS symptoms (Basnyat and Murdoch, 2003). Aspirin and Ibuprofen have also been shown to provide some beneficial effects.

Guidance should be sought from a general practitioner on altitude medication.

High-Altitude Cerebral Oedema (HACE)

High-altitude cerebral oedema is life-threatening. It is thought to be an extension of AMS, with much worse brain swelling. Those who have a 'tight fit' brain may be more susceptible to HACE as they experience an increased pressure when cerebral blood flow is increased. The incidence of HACE is difficult to ascertain, as it may be problematic to differentiate between AMS and HACE in some instances.

The symptoms of benign AMS progress with the appearance of ataxia (upset in gait characteristics), irrational behaviour and changes in the level of consciousness. There may be blurring of vision. Coma develops with irregular breathing, and death occurs within a few hours to a day or two.

High-Altitude Pulmonary Oedema (HAPE)

High-altitude pulmonary oedema is life-threatening. An increase in the pulmonary artery pressure (PAP) is an undesirable consequence of exposure to altitude, and is associated with HAPE. Fluid in the lung limits gas exchange. Possible contributory factors include exposure to cold and participation in high-intensity exercise, and a recent respiratory infection increases the risk. Those who have had HAPE appear to be more susceptible to further bouts, and are likely to experience HAPE again when exposed to similar altitudes. HAPE is normally found at altitudes over 2,500m and usually presents when the individual has been between 2–4 days at altitude.

Symptoms

Typically, the first HAPE symptoms are shortness of breath at rest or on mild exertion, and fatigue. The ability to exercise is reduced to a level that is even lower than that expected at altitude. Chest pain with tachycardia (elevated heart rate) and tachypnoea (elevated respiratory rate) with crackles (noises detected by stethoscope) at the lung base are often found. A dry cough later progresses to sputum, which starts white and frothy, but eventually becomes tinged with blood. Over a few hours there is increasing respiratory distress, cyanosis (a bluish skin colour), bubbling respiration, coma and death. Not all of the symptoms outlined above may be present.

Avoidance and Treatment

Very often it is not possible to have the appropriate resources or medication to counter HAPE, so it is most important to try to avoid it.

Treatment of HAPE and HACE

- Avoidance is desirable.
- It is important to recognize HAPE early and to take immediate steps to prevent further deterioration.
- Subjects who have a history of HAPE should ascend slowly. They may be able to prevent HAPE by taking 20mg of slow-release Nifedipine every 8 hours (Basnyat and Murdoch, 2003).
- The patient must be got down as fast and as low as possible, but this may be difficult in hostile terrain with a subject who is unable to maintain coordinated movements, and is fatigued. If possible, it may be prudent to carry the HACE victim down to a lower altitude. Immediate descent of even as little as 300m can produce a marked improvement.
- If available, a portable pressure chamber may be life-saving. A Gamow bag can be used if descent is not possible – the subject lies in the bag and the pressure in the bag is increased, using a foot pump, thereby 'taking the subject downhill'. (He may find it difficult to cope with the body position in the bag.) The bag should only be used to buy time to allow actual descent to take place. It also has the problem of being labour-intensive, and prolonged use may exhaust other members of the party.
- The use of 10mg Nifedipine and a follow-up of 20–30mg of slow-release Nifedipine, allied with descent and oxygen, have been recommended for HAPE.
- If available, oxygen should be given at a rate of 6–10L/min for the first few hours.
- Diuretics (to increase urine output) may be administered – Frusemide up to 1,000mg till diuresis.
- Morphine (10–15mg iv) may also be given.
- For HACE, 8mg of Dexamethasone has been shown to decrease symptoms but does not enhance performance in some physiological and psychological tests. Nifedapine, 20mg sublingually (under the tongue), may also be considered.

CHAPTER 7

Pollution

Pollution can adversely affect the athlete's health and performance, and lead to respiratory discomfort and various illnesses. All athletes should be familiar with the problems that may arise from pollution, whether at home or abroad. They should be aware of the strategies to minimize the difficulties that may arise, and should have plans to reduce any possible detrimental effects on performance.

AIR POLLUTION

Man-made pollutants from cars and factories will be influenced by the prevailing weather, the time of year and time of day. For example, carbon monoxide (CO) levels peak with rush-hour traffic. Ozone formation is associated with sunlight and, consequently, ozone levels are highest during the afternoon in summer. Knowledge of the major pollutants will help the athlete evaluate the potential risks and highlight the need for precautions or actions to avoid problems. A detailed overview of air pollution is given in *Air Pollution, An Introduction*, by Jeremy Colls (E & FN Spon, 1997).

Air quality is influenced by a number of factors. Pollution is most frequent in industrial areas and large cities. A city that sits in a bowl underneath a mountain is particularly vulnerable to air pollution, as the air is often stagnant; where there is temperature inversion (air sinking towards the valley), air pollution can be very bad. Generally, there is a decrease in air temperature as height increases, resulting in air movement upwards. Rising hot air removes pollution to high altitude, and thus provides an 'escape' for polluted air. In some circumstances, a layer of cold air blocks the upward movement of warm air and traps the polluted air. Thus the air is unable to rise to higher levels, and this results in pollution. Cities notorious for high levels of air pollution include Los Angeles and Athens. While the athlete may not live in a particularly polluted area, there may be times when pollution is bad, or a specific location (such as a small valley next to a motorway) may be highly polluted.

There are a number of sources for information on air pollution, including press and television, weather forecasts, and websites (*see* page 171), some of which give detailed information on current air pollution as well as records for particular locations. These may be particularly helpful for the athlete, but websites are not available for all areas in the world.

Primary and Secondary Pollutants

A primary pollutant is a substance that has a direct effect on the environment, where there is little or no change in chemical composition of the substance. Examples of primary pollutants are carbon monoxide (CO), which comes mainly from exhaust fumes from cars, and sulphur dioxide (SO_2), which is formed with the burning of fossil fuels such as coal and oil; power stations, steel works and some domestic heating appliances are sources of SO_2.

A secondary pollutant is created when an innocuous substance combines with an existing pollutant to form a different pollutant. For example, ozone is formed as a result of the interaction of sunlight and pollutants.

Effects of Pollution on Health and Performance

Air pollution can adversely affect the health of sedentary individuals, depending on the concentration of the pollutant, the length of exposure to it, and the amount of air breathed. Athletes are exposed to the same concentrations as sedentary individuals, but they will breathe much larger volumes of air during training and competition, and are therefore particularly prone to problems relating to air pollution. They are often involved out of doors in exercise using large muscle groups. They move large volumes of air, often for long periods of time, and thus expose the respiratory system to risk. Breathing through the nose can remove almost all of some pollutants (for example, the highly soluble gas sulphur dioxide), but this strategy is not available to athletes who, during high-intensity exercise, need to breathe through the mouth. Also, the rate of air flow into the lungs is greater during high-intensity exercise, and pollutants are consequently carried deep into the lungs.

When considering the effect of air pollution on athletes, it is necessary to examine not only the pollutant concentration, but also the athlete's volume of air breathed out in one minute, and the exercise time. Three hours at rest in a given pollutant may have no impact, but a 3-hour run in the same environment may result in adverse consequences.

Measuring and Reporting the Impact of Air Pollution

Air pollution is often reported by giving the pollutant concentration in parts per million (ppm), in parts per billion (ppb), or microgrammes per cubic metre (mcg/cubic metre) (for example, ozone concentration). The concentration of carbon monoxide (CO) in the blood is given as a percentage of CO combined with the haemoglobin (carboxyhaemoglobin, or COHb).

When studies on air pollution are interpreted, it is useful to compare the concentrations reported in the study with the levels that are likely to be experienced in the environment. For example, an individual sitting in an open-top car for 90 minutes on a Los Angeles road showed a level of 5.1 per cent COHb – that concentration is enough to decrease aerobic performance. The value would probably be even higher in an athlete who was running alongside the same road.

Lung function tests measure forced vital capacity (FVC), defined as the greatest volume of air that can be exhaled in one maximum expiration after a maximal inspiration. FEV_1 refers to the amount of air the lungs can exhale in one second after a maximal inspiration, and indicates how well the conductive airways are functioning. A reduced FEV_1 is an indication that the diameter of the tubes (bronchioles) that carry air to the alveoli is reduced and that air flow is restricted. This is one of the tests that can be used to assess whether air pollution has had an impact on lung function.

Air Pollutants

Carbon Monoxide (CO)

CO results from incomplete combustion, and is a fairly frequent cause of death in poorly ventilated houses with malfunctioning heating systems. CO is an odourless gas that is not detectable by humans without scientific apparatus. Thus, subjects are unaware of its presence. The symptoms and effects of CO vary, depending on time and severity of the exposure to CO. Symptoms from low carboxyhaemoglobin (COHb) levels may be a headache but, with longer exposure time to CO and increasing COHb levels, symptoms worsen, and include visual impairment, nausea and coma, and death can follow. COHb levels above 4.5 per cent can disrupt coordination.

Haemoglobin (Hb) combines with CO to form COHb around 230 times faster than it combines with O_2. CO blocks O_2 from combining with haemoglobin. In those haemoglobin sites where no CO has been bound on a specific haemoglobin molecule, it becomes more difficult for the haemoglobin to release O_2. This helps to explain why exercise that relies on the O_2 transport system may be impaired by CO inhalation.

In his review, Pandolf (1986) concluded that during submaximal exercise – 30–75 per cent of the maximal oxygen uptake (VO_2max) – COHb concentrations between 11 and 20 per cent (much higher than those normally found in the environment) have been reported to have no major effect on physiological responses. However, heart rate and minute ventilation (VE) were higher with increased COHb levels.

VO_2max is adversely affected by an increased COHb. A level of 4.3 per cent is the threshold for a decrease in VO_2max (Horvath et al., 1975), and a COHb above 2.7 per cent (Folinsbee and Raven, 2001) is the threshold for a decrease in performance during high-intensity treadmill exercise. Thus, the endurance athlete is likely to experience a decrease in performance with COHb levels above 2.7 per cent. Above these thresholds, the higher the COHb concentration, the greater the impact on exercise performance and VO_2max. Based on extrapolations from the literature, a COHb of 5 per cent will reduce VO_2max by 5 per cent, and a COHb value of 15 per cent will cause a 15 per cent fall in VO_2max. Such COHb levels can be attained by running adjacent to traffic and/or by simply sitting in a car in heavy traffic.

The mean CO level in the ambient air may not be particularly high over a long period of time. Carlisle and Sharp (2001) refer to the UK National Monitoring Network (UKNMN) when they highlight that reported CO levels rarely exceed 10ppm when averaged over an 8-hour period. However, Horner (1998) cites specific locations in south-west London where concentrations were higher: 13.8ppm (range 2–31) in a car, 44.0ppm (3–155) in an underground car park, and 10.9ppm (6–21) in a pub. Clearly, prolonged exposure to cigarette smoke can significantly raise per cent COHb. Horner considers that cigarette smoke was the main cause of the high COHb levels in the pub, and cites the work of Sterling et al. (1982), who reported CO concentrations of 0 to 5ppm in non-smoking areas in restaurants and nightclubs, compared with levels of 1 to 25ppm in smoking areas.

Gong and Krishnareddy (1995) reported that runners and cyclists involved in traffic jams had COHb levels of 4–6 per cent. Smokers who smoke between 10 and 20 cigarettes per day had COHb levels between 5 per cent and 9 per cent. Runners who ran for half an hour along a congested New York street had a COHb level of 5.1 per cent – similar to the levels of the smokers. After a 30-minute run on a road adjacent to Central Park, COHb levels were 4.2 per cent for the runners compared with 2.4 per cent for those who stood at the roadside for 30 minutes (Nicholson and Case, 1983). Horner (1998) cites other studies when he quotes CO levels of 9.8ppm in a car on a Californian road, 12ppm in a car in the middle of Paris and 56.1ppm in vehicles in Mexico City. The COHb levels of motorists involved in heavy traffic in Los Angeles increased from 1.12 per cent to 5.08 per cent over a 90-minute period (Hage, 1982).

Although CO emissions from cars have been reduced as a result of the introduction of catalytic converters, the overall CO emissions have remained relatively constant as the number of motor vehicles has increased (Carlisle and Sharp, 2001).

Given the COHb levels associated with decrements in aerobic performance, the athlete should be aware that exposure to certain environments may result in him under-performing. Those who are exposed to a tobacco-smoke-filled environment, or to traffic, and/or unwittingly spend time in an area with a high CO concentration, should consider the possibility of an adverse effect on performance.

Those in confined spaces should be aware of the risks from incomplete combustion. CO can cause death in a poorly ventilated house with a malfunctioning heating system. The risk may also apply to those at altitude using a stove inside a tent or enclosed space. At altitude any reduction in the oxygen-carrying capacity could have disastrous effects on performance and may, indirectly as well as directly, cause death.

CO and performance

- Carboxyhaemoglobin (COHb) levels of 2.7 per cent and above produce decrements in performance during high-intensity exercise.
- Exposure to a smoke-filled environment, traffic, and/or an area with a high carbon monoxide (CO) concentration may lead to an athlete under-performing. Travelling by car to competition and encountering traffic congestion, or travelling in a car with smokers will exacerbate the situation.
- Vehicle exhaust fumes are a major source of CO and should be avoided.

Sulphur Dioxide (SO_2)

SO_2 can cause damage to the airways and alveoli. As a considerable amount of SO_2 can be removed from the inspired air by the nose, it is possible that little or no SO_2 reaches the lungs during quiet breathing at rest. However, during even moderate-intensity exercise, it is not usually possible to breathe solely through the nose, and large volumes of air are often breathed in through the mouth. Consequently, a considerable amount of SO_2 can reach the lungs. SO_2 inhalation results in bronchoconstriction (contraction of the smooth muscles of the conducting airways) and restriction of air flow.

Unusually high SO_2 concentrations (1–2ppm) are needed to induce bronchoconsriction in normal subjects. These SO_2 levels are only likely in industrial districts. Asthmatics are nearly ten times more sensitive to SO_2 than non-asthmatics. Even short exposure times to SO_2 of about 0.25ppm and above will result in bronchospasm and a worsening of asthmatic symptoms (Gong and Krishnareddy, 1995). Symptoms for the exercising asthmatic in a SO_2 environmemt include breathlessness and a 'tight chest'. A combination of factors, including a switch to breathing through the mouth, increases in the volume of air breathed out in one minute, and a cold dry environment, are likely to exacerbate the bronchoconstrictor effect of SO_2 in asthmatics.

No studies have investigated the effect of SO_2 on maximal exercise performance. In his review, Pandolf (1986) concludes that the threshold for a SO_2 effect on humans during submaximal exercise lies at concentrations between 1.0 and 3.0ppm. While decrements in lung function in healthy subjects are likely to occur during light exercise with SO_2 concentrations between 1.0 and 3.0ppm, asthmatics can expect to have decreases in lung function at much lower SO_2 concentrations. Impairment of lung function during submaximal exercise is to be expected with SO_2 concentrations between 0.20 and 0.50ppm in asthmatics (cited in Linn et al., 1983). Nasal breathing lessens the increase in airway resistance but this option is not possible during high-intensity exercise.

Carlisle and Sharp (2001) consider that SO_2 values in the UK are unlikely to present problems for athletes with healthy lungs. They cite a SO_2 value of 0.1ppm as the recommended concentration for a 15-minute period in the UK. However, many athletes are asthmatic and are vulnerable to the adverse effects of SO_2, with the impact being greater in cold and dry compared with hot and humid conditions.

Limited evidence suggests that individuals may adapt to SO_2. Some subjects who have been exposed to SO_2 had less marked airway resistance in subsequent exposures. However, there are no studies that have investigated SO_2 adaptation and exercise responses.

Nitrogen Dioxide (NO_2)

Nitrogen oxides result from combustion of car or aircraft engines and emissions from power stations. NO_2 is much more toxic than nitrogen monoxide (NO), although there is a chemical equilibrium between them, with ozone (O_3) as the reagent. NO_2 is soluble and absorbed by the mucous membranes of the respiratory tract and converted to nitric and nitrous acids. At concentrations below 500ppb, it can impair resistance to respiratory infections, while acute exposure to 5,000–10,000ppb can result in coughing and breathlessness,

Prolonged exposure to 16ppb results in an increased risk of lower respiratory tract infections in children. It has been suggested that those with respiratory problems such as asthmatics may be particularly prone to the detrimental effects of NO_2, with airway resistance increases with exposure to NO_2 of around 500ppb.

A small number of studies have evaluated the effect of NO_2 during low- to moderate-intensity submaximal exercise. Typical, NO_2 concentrations were between 1–2ppm with exercise (intermittent or continuous) of around 2 hours. In his review, Pandolf (1986) concluded that NO_2 has apparently no adverse impact on physiological variables in healthy humans during submaximal exercise. Carlisle and Sharp (2001) cite the UK National Monitoring Network when they state that hourly mean NO_2 levels are usually below 0.15ppm in cities.

Comparison of the relevant studies suggests that NO_2 is not a problem for athletes in the UK. It is acknowledged that the small number of studies have been characterized by low minute ventilation (VE).

As with most air pollutants, traffic is the main source of NO_2, so high levels of NO_2 are related to traffic density, which is usually found in urban conurbations. While there are potentially serious effects, outdoor levels are usually low, and there can be greater dangers indoors in poorly ventilated rooms with gas or open fires.

Ozone (O_3)

Increased O_3 concentrations produce respiratory tract problems, and symptoms including coughing, chest discomfort, breathlessness and headache. Exposure to O_3 has resulted in a change in breathing pattern with an increase in respiratory rate and decrease in tidal volume compared with clean air. This change in breathing pattern is thought to be linked to the respiratory discomfort associated with O_3.

Folinsbee and Raven (2001) in their review highlight that possible adverse consequences of O_3 exposure include respiratory discomfort/chest pain, impaired lung function, reduced VO_2max (6–16 per cent), and reduced endurance performance (16–30 per cent). The impact of O_3 is related to the O_3 concentration, the time duration of exposure to O_3, and the VE. Impairment of lung function has been reported with exposure to O_3 levels of 0.2 to 0.5ppm for some hours at rest (Folinsbee and Raven, 2001). Such impairment is exacerbated by high ambient temperatures.

Endurance performance of around an hour has been dramatically decreased with increasing O_3 concentration. Ten endurance-trained cyclists performed four competitive cycle trials in the laboratory, breathing filtered air, and O_3 concentrations of 0.12, 0.18 and 0.24ppm. All subjects were able to complete the filtered air trial while nine finished the 0.12ppm trial, five the 0.18 trial, and only three the 0.24ppm trial (Schelegle and Adams, 1986). Decrements in FVC and FEV_1 were also reported after the 0.18 and 0.24ppm exposures. In addition, at the 0.18 and 0.24ppm concentrations, seven out of ten of the subjects reported a cough as a symptom, and eight out ten reported shortness of breath at 0.24ppm. Schelegle and Adams (1986) emphasized that international athletes who could sustain a VE greater than 100L/min for longer than an hour would probably be susceptible to the adverse effects of O_3 at

Possible adverse effects of O_3

O_3 concentrations of about 0.18ppm are likely to reduce elite endurance performance, promote respiratory distress and result in impaired lung function, particularly in athletes with high volumes of air breathed out in one minute over long periods of time.

concentrations under 0.18ppm. (O_3 concentrations above 0.20ppm are regularly recorded in Los Angeles in summer.) The reduced endurance performance has been attributed to respiratory distress. Impaired lung function and chest discomfort were features of the O_3 exposure. Foxcroft and Adams (1986) showed that VO_2max was reduced from 3.85L/min to 3.62L/min after 0.35ppm O_3 exposure for 50 minutes at a VE of 60L/min. Following O_3 exposure, endurance time was also reduced from 253 to 211 seconds.

In some environments, O_3 levels may be high for several consecutive days. Performance in lung function tests is reduced on initial exposure to O_3, with the decrement in lung function being worse on Day 2 than on the first day, but with an improvement on subsequent days. A lessening of the lung function impairment, and reduced O_3 symptoms, have been shown after 4 days of exposure to 0.40ppm O_3, although one study showed no improvement. Thus, individuals can adapt to O_3 but the degree of adaptation is related to a number of factors. Individuals who are very sensitive to O_3 take longer to adapt, and they lose their adaptation faster. Adaptation to a low level of O_3 will not provide adaptation to the same extent at higher O_3 concentrations. Adaptation can last up to 5 days (Florida-James et al., 2004). Some authors question the use of the word 'adaptation', and suggest that 'habituation' or 'desensitization' may be more appropriate. There is individual variation in the response to O_3, so it is difficult to predict how any one individual will respond.

Eight aerobically trained subjects were repeatedly exposed to 0.35ppm O_3 (4 days of 50 minutes' exercise at a VE of 60L/min). VO_2max and performance time increased (possibly due to a decrease in respiratory discomfort), but there was no concomitant improvement in lung function over the 4 days of the study (Foxcroft and Adams, 1986). Seven subjects reported that they had no subjective limitation on how hard they could exercise following 4 days of exposure, compared with only three subjects after the first exposure. While VO_2max and performance time can be enhanced with repeated exposure to O_3, repeated exposures could have undesirable health effects, including damage to the respiratory tract.

Carlisle and Sharp (2001) refer to the literature when they say that the daily ambient O_3 levels in the UK are not normally above 0.015ppm. In their review, they consider that these levels are not sufficiently high to present problems to athletes. Florida-James et al. (2004) point out that high O_3 values in Europe have been associated with exceptionally high summer temperatures. The year 2003 is given as an example. In Greece, O_3 concentrations were over 180mcg.m^3 (0.09ppm) for 94 days with a maximum value of 302mcg.m^3 (0.15ppm) being recorded. Florida-James et al. (2004) state that in the UK an O_3 concentration of 249mcg.m^3 (0.13ppm) was recorded on 37 occasions. With the rise in summer temperatures, it is possible that European O_3 levels will increase in the future. Peak O_3 levels of 0.54 to 0.60ppm and monthly levels of 0.23ppm have been recorded in Los Angeles (Armstrong, 2000; Foxcoft and Adams, 1986). Gong (1992) emphasizes that, while O_3 levels may peak in Los Angeles at about 0.35ppm, O_3 levels may remain above 0.12ppm for 10 days. He also adds that indoor O_3 levels may be 20–80 per cent of the outdoor levels.

Particulates (Microscopic Particles)

Particulates are a major component of fog. The winter 'pea-souper' fogs of the past were a combination of sulphur dioxide and soot, and occurred during high pressure with still air and temperature inversion. The cold moist air produced an aerosol of sulphuric acid and dirt, and the penetration into the lungs increased during exercise. The problem has been relieved in the UK with the Clean Air Act, but the soot has been replaced by diesel fumes. There may still be major problems with particulates in parts of the developing world and Eastern Europe, where they continue to burn soft brown coal, which not only produces half the energy of hard black coal, but also has a high ash and sulphur content. The problem is compounded by inadequate pollution control and poorly maintained diesel engines.

Summer smog is a cocktail of volatile hydrocarbons, oxides of nitrogen, sulphur dioxide, carbon monoxide and ozone, along with particulate matter. Again, as in winter, it occurs during a period of high pressure with temperature inversion. It is often worse in rural areas because there is likely to be more penetration by sunlight to increase the production of ozone. However, in urban areas with greater traffic pollution, nitrogen monoxide (NO) scavenges one oxygen molecule from ozone (O_3), producing nitrogen dioxide (NO_2) and oxygen (O_2). This is the normal chemical balance; *see* page 117.

Strategies to Minimize the Effects of Air Pollution

Any athlete who intends to compete in a location where there may be air pollution problems should make a risk assessment. There are websites that give information on current pollution problems, but many areas of the developing world are not covered. The websites allow scrutiny of previous air quality records and give an indication of possible difficulties by referring to records from previous years. Current values and forecasts can also be accessed. Based on such information, a strategy can be devised to limit the impact of pollution on health. Asthmatic athletes need to take particular care (*see* Chapter 9).

The athlete should determine when pollution is low and train at this time. For example, CO levels are generally highest during the morning and evening rush hours. Thus, training in the early afternoon is one way to avoid high CO exposure but, unfortunately, O_3 is highest in the afternoon – attempting to avoid one pollutant may result in problems with another. Athletes should avoid locations where CO is highest – next to roads and particularly in valleys, where the air is still – and train instead on hills or ridges, where there is likely to be more air movement and the pollutant will be removed. In some locations, it may not be possible to exercise away from traffic; in this case, the effect of the exhaust fumes may be limited by running or cycling on the upwind side so that the fumes are not swept into the respiratory system.

Training indoors can decrease exposure to air pollution, but in some situations it is still possible to experience air pollution; in one example, ice hockey players experienced high CO levels when the ice-smoothing system at the rink was faulty. Prolonged periods in traffic jams on the way to a competition may increase carboxyhaemoglobin (COHb) levels and result in impaired performance.

Exposure time to any pollutant can be reduced by arriving as late as possible into the polluted area, restricting the warm-up, and keeping car windows closed.

Minimizing the effects of air pollution

- Be aware that air pollution can adversely affect health and inhibit performance.
- Pollution may be very high in specific locations such as Los Angeles.
- Use as many sources of information as possible.
- Be particularly alert when calm weather conditions are forecast.
- Choose a hotel on a hill, not in a valley.
- Avoid air pollution as much as possible – stay away from heavily congested roads; train well away from cars.
- Wind promotes air circulation and reduces pollution.
- Ozone levels are usually highest in the afternoon.
- High carbon monoxide (CO) levels are associated with morning and evening rush hours, so avoid these times.
- Reduce the intensity and duration of training when air pollution is high.
- Train indoors but make sure that indoor pollution is not high.
- Avoid environments polluted by tobacco smoke.
- In competitive situations, reduce the time spent in the polluted area; delay travel to the area for as long as possible and keep car windows closed during travel.
- Reduce the extent of the warm-up.

WATER POLLUTION

Drinking Water

Water is unsafe to drink in major areas of the world because of bacterial contamination (Dawood, 2002). Even when drinking water is piped from deep sources it may still have dangerous mineral contaminants: lead has been found in the water in Mexico City (Neumann, 1999), and arsenic in Dhaka in Bangladesh (Leavitt & Vanarase, 2005). However, there may be isolated sources of pure water – one hotel in Dhaka had its own purifying plant and local medical and consular staff said that the purity was better than the bottled water. Even in areas where the drinking water is safe, it may contain chemicals, either naturally occurring or introduced as part of the purification process. These chemicals may cause stomach upset and should be avoided by anyone new to the area.

The advice is to use bottled water whenever possible, in plastic bottles with the seal unbroken. Carbonated bottled water is safer than still. Ice cubes must be treated with suspicion, because they are often made from unsafe water. Teeth should be cleaned in bottled water. Canned drinks are also generally safe, although there was a report in a Madagascan newspaper of a death that occurred through drinking from a can. This can had been stored in the refrigerator of a river boat and its top was badly contaminated with leptospirosis from rats' urine. Drinking directly from the can had transmitted the infection. The lesson is to decant drinks from a can into a clean container before drinking.

For those participating in sports where it is not possible to rely on bottled water (for example, mountaineering), it is possible to purify water by a number of methods (*see* Dawood, 2002). The simplest is to boil the water for 10 minutes.

Direct Contact with Contaminated Water

River water can be contaminated with a variety of toxic chemicals, including heavy metals, as well as with sewage. Similarly the sea may be polluted because of the outflow of sewage and drainage channels. A swimmer may ingest these toxins and infections, and infections with Hepatitis A virus have occurred due to swimming along a coastline with sewage outflow pipes, while Weil's disease has affected river canoeists. Recreational swimming in fresh water in certain tropical areas also exposes the participant to schistosomiasis infection through skin penetration. These swimmers may develop 'bather's itch', intense discomfort caused by the schistosomiasis organisms penetrating the skin. Swimmers in tropical waters may also rarely develop 'seabather's eruption', an intensely itchy rash limited to areas of the body covered by swimwear and caused by contact with the larvae of jellyfish, corals, sea anemones and hydra (MacSween & Williams, 1996). *See* also Chapter 11.

There is also a risk of neurotoxicity (damage to the brain and nervous system) in conditions where an algal bloom has occurred, and this may be in sea or fresh water (Dawood, 2002). (In an algal bloom, there is a sudden overgrowth of toxic algae, which is usually sufficient to cause the water to be very coloured – blue-green or orange. This should not be confused with the normal brown colour of muddy water, or water that has been running through peat.)

NOISE POLLUTION

Many parts of the world are very noisy, especially in built-up areas. Noise will disturb sleep, especially if it is intermittent, and traffic noise is worst in this respect. It is important for competing athletes to try and stay in a quiet room or use ear plugs to exclude the noise.

CHAPTER 8
Travel

Athletes often travel to other countries to compete, train, heat-acclimatize and take part in altitude acclimatization. A number of travel-related issues should be considered, including travel fatigue, jet lag, sleep deprivation, deep vein thrombosis, infections and the need for vaccinations to counter a number of diseases (*see* Chapter 10), contaminated food, polluted drinking water and gastrointestinal problems.

PRIOR TO TRAVEL

Health considerations – for example, the need for vaccinations (*see* below; Dawood, 2002; and websites) – for the proposed destination should be examined and acted upon, and the appropriate timescale should be determined. Prescribed preventive medication and potential remedies for likely occurrences should be sought (*see* below, and Dawood, 2002). For example, it may be worth taking some medication to combat diarrhoea.

The travel itinerary should be written out, with check-in times highlighted. Travel documents should be stored in a zip pocket. The travel checklist should include travel tickets, information on how to reach the hotel from the airport, hotel details, passport, visas, and vaccination information. Verification that luggage meets the airline regulations should be made at home and not at check-in. Travel arrangements to the airport should allow plenty of time with something in reserve to cope with emergencies. Similar processes should be adopted when travel is by means other than plane.

When travelling abroad, adequate medical insurance is essential and any policy must include the cost of flying the person home, when appropriate, due to illness or injury. It is sometimes possible to obtain treatment free or at reduced cost when abroad. Some countries have reciprocal health agreements with the UK and the European Health Insurance Card will access that within the European Community and Switzerland. Sometimes money has to be paid out first and then reclaimed in the visited country. In some places consultations may be free but not medication, so there must be sufficient funds available for difficult situations. Leaflet T4 from the Department of Health provides the necessary information and European Health Insurance Card application.

PROBLEMS DURING TRAVEL

Motion Sickness

The vehicle used during travel – plane, ship, train, bus or car – may have many different movements that are not experienced by the body during 'normal' activity. The brain receives signals from the eyes, the vestibular (balance) mechanism in the middle ear and from the effects of gravity putting pressure on the part of the body in contact with the ground (or chair, or bed). In normal activity the messages received from the three systems are in agreement. However, when the signals from the sensors are not in agreement, the brain becomes confused about the exact position of the body in space and motion sickness develops. This typically occurs on a ship in rough weather. Individual susceptibility varies greatly but no form of travel is exempt. While motion sickness is a temporary phenomenon, it can be devastating and leave the sufferer feeling very weak for several hours after the motion has stopped.

Prevention is better and easier than cure and anyone who knows, or thinks, that he is susceptible should take oral medication at least an hour before travelling. If an athlete has not had the opportunity to ascertain his susceptibility to motion sickness in general, or to the particular form of travel being used, he might find it advisable to take precautions anyway, especially if there is limited time available after arrival.

There are a variety of medications available, but the following are the most common:

- hyoscine oral (Kwells): 0.3–0.6mg 30 minutes before, and repeated every 4 hours;
- hyoscine patch (Scopoderm): one patch behind ear 6–8 hours before; lasts 72 hours;
- cinnarizine oral (Stugeron): 15–30mg 1–4 hours before and repeated every 8 hours; cinnarizine is frequently prescribed because it has less sedative effects than other antihistamines.

Oral tablets should be taken regularly as long as the travel or motion continues. Once a person is sick, oral medication is ineffective because, even if the tablets are not vomited up, the stomach is in stasis and the chemical will not be absorbed. At that stage an intramuscular injection will be needed. A frequently used one is promethazine (Phenergan) 25mg. This will be effective in about 15 minutes and the person will probably fall asleep. It may last up to 18 hours but after 3–4 hours, when the person has woken up, oral agents should be started before the promethazine has worn off.

Deep Vein Thrombosis (DVT)

DVT describes the condition where blood clots develop in the deep veins, usually in the calf. This can result in swollen, uncomfortable legs, which will obviously impair performance, but the major danger is if these clots become loose and travel up the veins to the lungs, causing a pulmonary embolism (blockage in the blood vessels in the lungs). If this happens, it can be immediately fatal.

DVT is frequently called 'economy class syndrome' but in fact it can occur in any situation involving prolonged sitting, especially in cramped conditions, including a long car or bus journey, as well as sitting in a plane – in any class.

Those at greatest risk of DVT include smokers (particularly those over 40 years old), anyone with a history of blood clots in legs, or previous surgery, and women taking the oral contraceptive pill.

Deep sleep may contribute to the incidence of DVT through lack of movement, so sleeping pills should be avoided. Regular bouts of exercise throughout the journey, particularly using calf muscles (for example, ankle circling), will lessen the chances of contracting deep vein thrombosis. Regular walks to collect and take on fluids, look out of windows and visit the toilet will help avoid stiffness. In the seat, heel raises and knee extensions should also be an integral part of the exercise routine.

Prevention of DVT

- Avoid tight clothing round the legs, such as ill-fitting socks or tracksuit bottoms.
- Wear elastic compression stockings during the flight.
- Take aspirin 300mg immediately before journey.
- Deep sleep may contribute through lack of movement, so avoid sleeping pills.
- Exercise regularly throughout the journey.

Dehydration

During a flight, the dry cabin air promotes respiratory fluid loss and dehydration. Water and/or fruit juice should be consumed. Alcohol, tea and coffee result in increased urine production and are likely to promote dehydration. Dehydration and the lack of fibre in the food offered on aircraft may combine to promote constipation, so the traveller should make extra efforts to drink water and fruit juice regularly, and to eat fresh fruit during the trip.

TRAVEL FATIGUE AND JET LAG

Long-distance travel by any form of transport is likely to result in travel fatigue. Disruption to normal lifestyle, monotony, dehydration and travel stress may all contribute to a feeling of overwhelming tiredness. Sound planning has the potential to reduce fatigue. Finances may dictate the travel timetable, but early rises and long stopovers should be avoided when possible. Currently, flying is the most popular means of long-distance travel and this introduces its own problems. Flights that involve east–west travel will pass through different time zones, and are likely to result in jet lag as well as travel fatigue. Travel fatigue is likely, however, even if there is no jet lag, for example, on a north–south flight from the UK to Africa. The choice of a night or daylight flight should be considered (*see* below). Early bookings may enable the traveller to procure a 'long leg seat'.

North–South Travel – Latitude Crossing

With north–south travel there is no jet lag because the traveller remains in the same time zone, or crosses only one or two zones. However with latitude change there can be variation in hours of daylight and length of twilight. The effect is related to movement nearer to or away from the Equator. Moving a significant distance away from the Equator in summer, there is a great increase in daylight hours; this causes some travellers to become very short of sleep – they cannot go to sleep until darkness falls, and wake up again as soon as it is daylight. These people have to be very careful to sleep in rooms where daylight can be excluded.

In winter the reverse is the case, with very little or no daylight. This can cause problems in waking up. It will also reduce the time for daylight training. Some people are so badly affected by lack of daylight that, if exposure is prolonged, they may develop Seasonal Affective Disorder (SAD) syndrome, a form of depression.

Moving nearer the Equator the hours of daylight become more regular throughout the year, closer to 12 hours of daylight and 12 hours of night. However, the period of twilight shortens, with the sun setting more vertically. If an athlete from a high-latitude country is used to going for a long run in the evening as the sun is approaching the horizon, he could be badly caught out in the tropics. In the higher latitude, there may be 3 hours of gradually darkening twilight; in the tropics, sunset to total darkness may last just 10 minutes, and the athlete may find himself in total darkness, and totally lost.

East–West Travel – Longitude Crossing

There are 24 time zones in the world and local times are related to the position of the sun at a given location. At noon in London, it is 7am in New York and 4am in Los Angeles; these cities west of London are behind London time. Beijing to the east of London is 8 hours ahead. Time zones follow country borders rather than having a direct north–south boundary. For example, Portugal is on GMT but over the border due north in Spain, the time moves forward one hour because Spain is on European time (GMT + 1 hour). Continuing due north, the traveller will reach Britain and have to put the clock back 1 hour again into GMT.

Biological Rhythms

The body has a daily cycle (circadian rhythm) for a number of functions. This biological clock enhances mental and physical functioning during the day and promotes sleep and recovery at night. Biological rhythms influence many of the body's physiological variables, including core temperature, heart rate and blood pressure. For those who sleep at night, values for these variables are at their lowest towards the end of the night, and at their peak 12 hours later, whereas many hormone concentrations are higher at night. Evidence of the existence of this body clock can be observed when travelling from the UK to the west coast of the USA, which is 8 hours behind UK time. The journey is tiring, and the individual is unlikely to have slept well in 24 hours, but it is common to wake up early in Los Angeles morning, at around 3am (11am UK time). Over several days the wake-up time is later, and after a week the individual is likely to be linked into local time.

Biological rhythms are influenced by internal and external factors. The body clock runs slightly slower than 24 hours and without external stimuli would operate at 25 hours per day, so there is a need for ongoing adjustment to allow synchronization with a 24-hour day. Evidence for an internal factor comes from a study that removed any environmental or lifestyle influence. Despite the lack of external influences, the body rhythms were maintained, albeit on a 25-hour cycle, confirming the notion of a 'biological or body clock'. Factors that help with the adjustment of biological rhythms are often called *Zeitgebers* (German for 'time-givers'), the most important being the light/dark sequence, and secretion of melatonin from the pineal gland (Reilly and Waterhouse, 2005). Melatonin is switched off by light and switched on by darkness.

With the advent of jet aircraft it is possible to travel through several time zones in a few hours. This travel upsets the normal body clock stability by producing a mismatch between the external (local) and internal (biological clock) time, resulting in the phenomenon known as jet lag. The actual time in the new location may be 9am but the 'body's time' may be 1am, the time at home. Travel through several time zones means that the athlete may need to compete and train at a time when he would normally be in bed asleep.

Symptoms and Effects of Jet Lag

There are many symptoms associated with jet lag:

- fatigue;
- inability to sleep at night;
- feeling sleepy during the day;
- reduced physical performance;
- gastrointestinal problems: blunted appetite; irregular bowel movements;
- psychological problems: irritability; emotional lability; feeling 'below par'; limited concentration; low vigour and motivation.

Travel to the west results in less severe symptoms compared with eastward travel, and the severity of the symptoms rises with the number of time zones crossed. Waterhouse et al. (2004) highlight that some jet lag symptoms were reduced faster than others. They monitored the jet lag symptoms of athletes who travelled from the UK to Australia (10 time zones) and found that reduced concentration and sleep irregularities took longer to subside, compared with a decreased appetite and an increased irritability.

Reilly and Waterhouse (2005) highlight that there is a lack of consensus on the standardization of symptoms, which results in problems of interpretation of studies. For example, the importance of a specific jet lag symptom, and when the symptoms are measured, can influence jet lag scores, and consequently how jet lag is studied and reported. Winget et al. (1985) reported that, after westward travel over 6 time zones, heart rate and sleep/wake rhythms had readapted after 2–3 days, whereas core temperature was more likely to need 5 days and bowel movements 8 days for readjustment.

Scrutiny of the symptoms of jet lag suggests that athletes are unlikely to perform as well when their body clock has been disrupted. For example, lowered concentration and motivation are likely to have an impact on performance. A reduced performance in training may have an adverse effect on morale, and lead to concerns that the weakness may extend to the competitive setting.

Greater performance decrements are to be expected with eastward travel. In their review, Winget et al. (1984) cite studies that show that performance on flight simulators and reaction time was more severely affected with eastward travel than with westward travel. Winget et al. (1985) conclude from the literature that adaptation to jet lag is 30–50 per cent faster with westward travel compared with eastward travel. They suggest that, as the body clock runs slightly slower than 24 hours without external stimuli, it is easier to adapt to a phase delay (westward travel).

Waterhouse et al. (2004) highlight in their review that performance decrements have been attributed to time zone changes in a number of activities. Sprint and middle-distance running performance was reduced after crossing six time zones in an easterly direction. Grip and leg strength were lower following time zone changes, and it was 4 to 5 days before peak scores returned to normal.

Timing of Events

While there may be some bias as a result of the timing of events, Atkinson and Reilly (1996) cite data that suggest that athletic performance for many activities is best in the late afternoon. They refer to cycling trials and swimming events as examples of sports that normally have best performances in the late afternoon. They concede that the time of scheduling the event, and the fact that summer afternoon temperatures are high, may combine to bias the frequency of high performance to late afternoon. There is also individual variation, and some athletes will perform better in the morning, while others may excel later in the day. As core temperature is highest in the early evening, long-distance runners may not perform as well in the evening. They will start the event with a 'high' core temperature and reach their core temperature ceiling sooner than they would with a morning start (*see* Chapter 2).

Many factors, including strength, hand and eye coordination, and motor skills, contribute to athletic performance. The fact that performance in reaction time, coordination and strength is best in the late afternoon, which coincides with the time of highest core temperature, supports the view that participants in those sports where these components are important are likely to perform best in the late afternoon. As a result athletes performing in different time zones from home may reach their peak performance at an inappropriate time of day. Thus, there is likely to be a need for the athlete to adapt to a new time zone in an attempt to optimize performance.

As the factors that influence performance in different sporting events vary, it is likely that the time of optimal athletic performance will not be the same for all events. Monitoring performance in training sessions may give the athlete information on optimal performance time.

TACTICS BEFORE TRAVEL

Driving long distances across time zones will give protection from jet lag, as the time taken for the journey will allow the athlete's body clock to adjust to the new time zone.

If the athlete's stay in the new time zone will be a short one, adaptation will not be possible. Consideration should be given to the synchronization of important events with times that correspond to daytime at home. The athlete who has travelled west should focus on morning activities, which equate to the afternoon or evening at home, while the eastward traveller should carry out important activities in the afternoon or evening, which relate to morning at home. It may not be possible to achieve this strategy if the competitive event is scheduled for a time that is inappropriate for the athlete. However, if team officials have any say in the timing of an event, the above strategy should be applied – for example, a UK soccer team invited to Japan for a one-off game should attempt to play the game in the evening local time.

Given the likely performance and training decrements as a result of jet lag, the athlete should examine strategies to help the body clock come into line with the new time zone as soon as possible. The direction of travel and the number of time zones should be considered. Flying east, for example, from the UK to China, the flight should allow for sleep throughout and arrival in China in the morning. The timing of the athlete's arrival has been shown to influence jet lag. A flight to Australia from the UK via Kuala Lumpur or Singapore normally takes about 24 hours. Waterhouse et al. (2002) compared subjects who arrived in Australia in the afternoon with those who arrived in the early morning (Australian time). They concluded that jet lag symptoms were less in those who arrived in the afternoon compared with those who reached their destination in the morning. The former had less sleep during their first flight than the latter. The authors concluded that the advantage for the afternoon arrival was that the travellers were able to have a full night in a comfortable bed 30 hours after they rose from their bed in the UK, whereas the morning arrivals had a 50-hour gap between these time points. Plans should be made so that the athlete is able to sleep in bed at night-time in the new location with a minimum possible delay from the previous night's sleep in a bed.

Adjusting lifestyle may be worthwhile. For example, for east–west travel, the athlete could stay up later at night, and get out of bed later. For west–east travel, up to 8 hours of a change in time zone, it is desirable to advance the body clock, perhaps by rising and going to bed early at home. Of course, it may not be possible to amend schedules due to work commitments or family considerations, and it may also prove to be of limited value because it will clash with the powerful *Zeitgeber* of the light/dark cycle, which may disrupt the planned adaptation. Winget et al. (1985) reported that this approach was unsatisfactory with athletes because of the associated effort.

ADAPTATION

Given time, the body adjusts to the new time zone, but different bodily functions adjust at different rates. Therefore precise times are difficult to prescribe.

The athlete should bear in mind that the greater the number of time zones, the longer adaptation will take. Adaptation is generally faster with westward travel.

During Travel

The watch should be set to the destination time as soon as possible and food and sleep should be taken as if at the destination. (One watch should be kept at current time throughout the journey to ensure that the correct connections are made.)

Some researchers advocate that travellers from the UK to the west of the USA should stay awake during the flight (when it is daylight). However, a nap may help the traveller to stay up more easily when he lands in the west of the USA and this will aid his adaptation to the new time zone.

On Arrival

It is important to fit into local time as soon as possible. The *Zeitgebers* must be strengthened, especially the sleep/wake cycle. This involves going to bed and getting up at normal local times, even if this means lying awake for 3 hours in the morning, and feeling very sleepy in the early evening. Meals should be eaten at local times. It is important to avoid the temptation to go to bed early. Strategies to help stay awake include staying outdoors in light, exercising in a group, and maintaining social contacts until it is time for bed at the bedtime of the new time zone.

Since performance is likely to be impaired as a result of jet lag, full-out training sessions in the early days in a new time zone may be counter-productive. Waterhouse et al. (2004) advocate that, in the first few days in the new time zone, athletes should train in the overlap time between the new and old time zones – in other words, the waking time that is common to the two time zones. They also recommend that players should play a friendly match near the end of the first week of adaptation to the new time zone. Athletes and players should be aware that, during the first week of adaptation, their motor skills may be temporarily impaired. This can reduce their belief and assurance in their ability, and may increase the likelihood of injury.

Melatonin

The effect of melatonin on the adverse effects of jet lag has been assessed in a number of studies, with several, but not all, showing a decrease in symptoms. The mechanism(s) relating to the possible beneficial effects on jet lag are unknown, but assistance with body clock changes and the promotion of sleep are possible candidates. In their review of strategies to decrease the effects of jet lag, Herxheimer and Waterhouse (2003) say that 2–5mg of melatonin, taken before retiring to bed in the new time zone, will reduce the effects of jet lag when allied to the other non-pharmacological methods outlined here. Waterhouse et al. (1997) state that melatonin should be taken in the early evening in the new time zone, regardless of the direction of travel and the magnitude of time zone difference. This is likely to reduce jet lag symptoms. However, if melatonin is to be used to aid with body clock adjustment, consideration must be given to the direction of travel and time of melatonin ingestion. Melatonin given in the afternoon/evening will advance the biological clock (eastward travel), but if it is given late at night or in the morning it will delay the clock (westward travel).

Herxheimer and Waterhouse (2003) and others add cautionary notes about the use of melatonin. Comparison and interpretation of studies that have examined the effects of melatonin are difficult because of methodological differences. For example, markers of jet lag have not been consistent in all studies. However, the majority of studies have reported that melatonin reduced jet lag symptoms.

With regards to safety, the quality of melatonin in some countries has been questioned, and other, unknown substances have been found in some batches. Herxheimer and Waterhouse (2003) highlight that, in some countries, melatonin is deemed to be a medicine, and as such requires a licence. However, melatonin is regularly sold without a licence.

While Reilly and Waterhouse (2005) conclude that melatonin ingestion is likely to reduce jet lag symptoms, enhance sleep and improve physical and mental performance, they caution that some aspects of melatonin use remain unexamined. For example, it is unclear whether there are long-term side effects.

Buscemi et al. (2004) carried out a systematic review of the literature on the effect of melatonin on various sleep disorders. They concluded that, in patients with jet lag, melatonin did not decrease sleep onset latency or increase sleep efficiency, sleep quality, or the time spent in REM sleep (*see* below), although it was effective in increasing the total sleep time. In one study they examined, it was found that, in jet lag, melatonin had a similar effect as the short-acting hypnotic zolpidem.

It has been advised that those who are on warfarin and oral anticoagulant medication or who are epileptic should not take melatonin (Herxheimer and Waterhouse, 2003). Athletes should consult their GP about the possible use of melatonin.

Rapid Eye Movement (REM) sleep

There are several different types of electrical activity during sleep. One pattern is concurrent with rapid movement of the eyes and these periods of sleep are therefore referred to as Rapid Eye Movement (REM) sleep. REM sleep spells are the most effective periods of sleep with regard to recovery during the night.

Other Aids to Adaptation

Hypnotics (sleeping tablets), particularly short-acting agents such as zopiclone (Zimovane) or zolpidem (Stilnoct), will assist the development of the correct sleep pattern. They should only be used in the first 2 days.

Alcohol induces slow-wave sleep, but it disturbs the more beneficial REM sleep, and its diuretic effect may lead to further disturbance. However, a single nightcap on the early nights may be helpful. Remember: alcohol is a banned substance during competition in some sports.

Strategies to lessen the effects of jet lag

- Determine the number of time zones involved and the direction of travel, and decide on an appropriate strategy to optimize adaptation, if it is necessary.
- Avoid sleep deprivation before travel.
- Consider some pre-adaptation at home by adjusting your sleep/wake cycle to correspond more closely with the new time zone.
- Assess which direction of travel is better when trips are over 10 time zones.
- When travelling, set your watch to the time of your final destination.
- Fit into the meal times, social activities and sleep/wake patterns of the new time zone immediately on arrival.
- For westward travel, make use of the afternoon and evening light to take light exercise or go sightseeing, in order to stay awake while it is light.
- Sleep in a room with windows, to help adaptation to the light and dark sequence of the new location.
- Exposure to outdoor daylight, and avoidance of light at appropriate times, are likely to enhance adjustment to a new time zone.
- Initial training sessions should take place during the overlap period in time zones to help maintain the quality (particularly skill-related factors) of the training sessions; this will help avoid a drop in the athlete's confidence.
- The use of melatonin can advance or delay the body clock depending on the time of ingestion, and has the potential to enhance adaptation to a new time zone.
- For some, melatonin use is deemed to be inappropriate.

Physical activity may also help the adjustment, however, training schedules may have to be altered. For example, a 'normal' afternoon nap should be abandoned if it happens to coincide with night back home, since this will be read as 'night' by the body and will delay adjustment.

SLEEP DEPRIVATION

Sleep deprivation is difficult to define, but individuals who have less sleep than their normal (usually around 7 hours) are likely to be susceptible to the adverse consequences of sleep deprivation. Armstrong (2000) in his review states that the impact of sleep deprivation is greater for cognitive function than for motor skills. However, these decrements tend to be small when sleep deprivation is under 54 consecutive hours. Unless there have been major travel delays, it is unlikely that athletes will experience such serious sleep disruption. Armstrong (2000) suggests that disruption to body rhythms resulting from jet lag is more likely to reduce athletic performance than any sleep deprivation that is likely to be encountered during air travel.

RISKS TO HEALTH WHEN TRAVELLING

Food

There are several potential problems from food when travelling, including the risk of infection from badly prepared food, or contamination by flies. Anecdotally, in developing countries, food prepared and cooked by local people in their own way is usually safe. Many outbreaks of food poisoning occur in five-star hotels where attempts are made to provide 'western' food, but storage and refrigeration methods are misapplied. Infections such as tapeworms or toxoplasmosis may originate in the food, or be introduced by dirty hands, or by washing in contaminated water.

Cold foods, salads, fillings and sauces should be avoided. Cooked food should be steaming hot when served, and it is safer to have a dish with small pieces of meat, since the food is more likely to be cooked through. Poultry is a common source of salmonella and campylobacter. Some tropical fish contain toxins, which can cause gastrointestinal and neurotoxic symptoms. Shellfish should be avoided, with mussels in particular being renowned for transmitting hepatitis A. Unpasteurized milk, including cream and yoghurt, is also a possible source of tuberculosis, brucellosis, Q fever, salmonellosis, and campylobacter infections.

Fresh fruit and vegetables must have been cleaned in uncontaminated water. A simple philosophy for fruit and vegetables is 'Peel it, cook it or bin it'.

Being in a new country, it is tempting to try exotic meals but this should be avoided until after any competition, because unfamiliar foods may cause gastric upset.

Drinking Water

Treat all water and ice cubes suspiciously, and use bottled water when possible. (*See* Chapter 7 for more on this subject.)

Hygiene

One of the most common routes for transmission of infection is the faecal–oral route and this applies for salmonella, E. coli, cholera and rotavirus. To break this transmission it is important to wash hands regularly with soap and water, especially after using the toilet and before meals. Additionally, an antiseptic preparation may be applied to hands before touching food. Wipes are often used, but they have to be pulled from a container and, being alcohol-based, dry very quickly. Also they tend to be rubbed only into the palms of the hands. A water-based foam, dispensed by plunger from a bottle, is much more effective. Since it evaporates less quickly, it has to be wiped thoroughly into the hands before it disperses, and it remains moist for long enough to have a bactericidal effect. The NewGenn Foam Hand Rub has passed the EN 1276 efficacy test and appears to be effective against viruses including noro virus, and the blood-borne viruses hepatitis B, hepatitis C, and HIV. (Further information is available from the manufacturer, *see* Further Reading and References.)

In remote areas – for example, at high altitude – it is important to ensure that disposal of human waste is distant from the camp. It is also important to take drinking water upstream from any area used for waste disposal, always remembering that one person's upstream may be downstream from someone else's waste disposal area. In these situations, personal hygiene is even more important than usual. Water purification should also be used (*see* Chapter 7).

Gastrointestinal Problems

Many gastrointestinal upsets are the result of bacterial or viral infections following a failure of hygiene, but some may be caused by chemicals. For example, drinking water may be 'safe' in terms of infections, but it may contain chemicals that are either in the system or have been used for sterilization. Any chemicals that are unfamiliar to the traveller may cause gastrointestinal problems. These problems are not dangerous to health but will impair sports performance, and spoil a holiday. *See also* Chapter 7.

Bites and Stings

Bites may be inflicted by animals, snakes, arthropods or insects, and an athlete may suffer a bite or sting during training, competition or leisure, on land, in fresh water or in the sea. It is worth checking on potential hazards when travelling to a particular destination. Bites may occur in temperate areas as well as in the tropics and, with climate change, some dangers are currently moving north. For example, tropical needle fish, normally only found in the Indo-Pacific region, have 'wiped out' a professional surfer in the waters off Portugal (Kerkhoffs et al., 2003), and mosquitoes have been reported as breeding in the USA. The frequency and bulk of goods and people being transported by air mean that dangerous species from one area may be transported to and thrive in a distant country; one example is the notorious Australian redback spider, which is now established in Japan (Horton, 1997). Some people living near international airports have developed even malaria from mosquitoes that have hitched a ride.

A bite or sting may impair performance in several ways:

- it may cause tissue damage;
- it may cause irritation if it is in a sensitive spot, for example, where a shoe or waist band is tight or rubbing;
- it may become infected, which will impair performance during exercise, not only if the bite is near a joint, but also from the toxic material, and through the effects of any antibiotic prescribed;
- it may transmit a systemic infection, such as malaria and Lyme disease, which may have serious consequences.

Animals

Most animals are more inclined to run away from than to approach strangers. However, animals infected with rabies seem to lose their fear of humans, and travellers should be wary of any dog, cat, fox or other animal that approaches. Anyone who is bitten by an animal that has approached voluntarily should seek treatment, on the assumption that the animal may have rabies (*see* Chapter 10).

Snakes

Similarly, most snakes will get out of the way of humans and will only attack if they are stood on while asleep, or if the person is approaching a nest or young. Death resulting from an encounter with a rattlesnake is almost invariably due to the snake being teased or bothered. No one should walk in undergrowth, or deep sand, without boots, socks and long trousers. Anyone approached by a snake should not run, but should stand still and allow the snake to pass. Snakes only strike at moving objects. *See also* Dawood (2002).

Treatment for snake bites consists of reassurance, pain relief and immobilization of the limb and the person to avoid spreading the toxin. A soft pad should be placed over the puncture mark, and a compression bandage applied from below upwards (Warrell, 2005), and medical or hospital advice should be sought.

Insects

Some insects bite humans in order to feed, and many diseases, including malaria and yellow fever, can be transmitted through such bites. The ideal combination is to prevent bites where possible, have adequate appropriate immunization, and use preventive drugs if available (*see* Chapter 10).

Scorpions are found in India, Africa, the Middle East, and North, Central and South America. Stings are very painful but not usually fatal in adults. In an area where scorpions live, always examine and shake out shoes, including running shoes, before putting them on.

Dangerous spiders are found in Australia, South Africa, the Mediterranean countries and North, Central and South America, although some species may be moving north. Spider bites are rarely life-threatening in adults.

Protection from Bites and Stings

Clothing should be made of tightly woven fabric (insects can bite through sheer fabrics and tight clothes), with a high neck, long sleeves and legs; a bandana may also be worn around the neck. If the insect problem is particularly bad, consider a hat with attached net over the face and neck. Material should be light coloured since insects are not attracted and, because they are easily seen, are easier to remove. Wear high socks and shoes.

Avoid perfumes, hair sprays, cosmetics and jewellery. At night, air conditioning allows windows to be kept shut and insects are less active because the environment is cool. If air conditioning is not available, beds should be draped with mosquito nets (at least 18 threads per inch). There should be adequate space between the net and the sleeping person, because insects will bite if the body is touching the net. The net must be tucked in under the mattress and must have no tears. Ideally there should be a zipped entrance into the net space.

Smoke and vapour from burning are ineffective in open well-ventilated areas and may cause respiratory problems following prolonged indoor exposure. Sound-making devices are ineffective, and light traps are counter-productive, attracting all insects but killing mainly harmless ones.

Skin-So-Soft, a concentrated bath oil, will offer some protection for 60–90 minutes.

The insecticide Permethrin rarely causes side effects (some users may experience mild skin rashes), and has no serious toxic effects. It is safe for all fabrics and should be sprayed around the bedroom (on the curtains and bed, as well as any netting) 30 minutes before retiring. Footwear, clothing and ankle bands may be impregnated with Permethrin for protection during the day.

Diethyl Toluamide 'Deet' is an insect repellent that has been used by the military for many years and is the main constituent of many products, including Jungle Formula and Repel. While there are some anxieties about Deet, these are probably exaggerated. Insect repellents that do not contain Deet include Mosiguard and Autan.

Insect repellents are available in the form of lotions, creams, roll-ons or sprays. The spray should be sprayed on the hands, the hands rubbed together, and then the preparation spread over the area to be covered, taking care round eyes and lips. Simple spraying tends to protect everything round the person, rather than the person himself.

MEDICATION FOR TRAVEL

It is impossible to provide a comprehensive list of medications for all venues. The list of medications that should be carried depends on the destination and the potential hazards in the area. It is also important to consider the availability and the standards of local medical support. For example, one of the commonest methods of transmission of hepatitis B, which is present in many areas of the world, is by inadequately sterilized surgical instruments or needles. Where there is any doubt, it would be advisable to carry a simple kit to suture a cut and not rely on a local medical centre. For further information, consult Dawood (2002), Auerbach (1995), and Pollard & Murdoch (2003).

Athletes competing in top-level competition – Olympic Games, World or Regional championships – must of course ensure that they are not using a substance banned by the IOC and WADA. (*See* the list of websites on pages 170–1, since the details change every year.)

CHAPTER 9

Pre-Existing Medical Conditions

Many medical conditions cause concern because of their implications for exercise and travel. For example, a previous heart attack might be considered a contraindication for running a marathon – although this is not in fact an absolute truth (Kavanagh, 1976). Various conditions are discussed here in the context of specific situations and difficult environments. For other conditions, and their relevance to travel, *see* Dawood (2002), Auerbach (1995) or Pollard & Murdoch (2003) page 184. Of course, everyone should also consult his or her doctor.

ATHLETES WITH DISABILITIES

Heat exposure may have specific dangers for some athletes. Within Les Autres group, the condition of some athletes may be temporarily or permanently exacerbated, for example, in multiple sclerosis. Heat also has the potential to affect muscle tone, and this may have profound, and unexpected, effects on athletes with cerebral palsy.

Dehydration may alter the effects of medication, particularly antimuscarinic drugs (which suppress the action of specific receptors, and are used in the treatment of diabetic neuropathy and asthma, as well as in other situations) and of medication given to control high blood pressure. The side effects of the changing environment on medication may hinder athletic performance, for example the visual impairment caused by diabetes.

Some athletes with disabilities have specific problems in regulating body temperature. For example, athletes with spinal cord injury (SCI) have a reduced sweating capacity of the insensate skin (which lacks feeling), and a lack of vasomotor regulation (control of the arterioles in response to temperature changes) below the level of the lesion (injury). Therefore the thermoregulatory system is compromised in terms of the available cooling mechanisms and blood redistribution during exercise. Even in a thermoneutral environment, an SCI individual has different thermoregulatory response at rest, during exercise and into recovery compared with an able-bodied individual. They are therefore at greater risk in a hot or humid environment. Athletes with bilateral amputations, especially if high, may experience similar difficulties, since they have a much smaller surface area for heat loss and for sweating.

The function of a shunt (a tube with a one-way valve inserted to allow excess fluid in the brain to be drained into the main circulation) may alter when exposed to the sun for a long time.

Sweating will increase problems with pressure areas particularly in wheelchair athletes and athletes with artificial limbs. Pressure areas must be given extra care

– for example, by more frequent changes of stump sock. It may be necessary to pad stump sockets with cotton wool or orthopaedic felt if there is shrinkage of the stump. Sweating may also lead to problems in skin folds – for example, in the groin or under the breast – and this may lead to fungal infections. Problems may also arise with athletes who use urine drainage bags, and absorbant material should be placed between the bag and the skin.

A change of time zone may lead to problems or worries about the timing of medications. Most of these can easily be sorted out by discussion with the medical team or GP.

The track surface may be so hot that repeated starting practice leads to burn-like discomfort in the fingertips. This can be reduced by athletes pouring some water along the start line where their fingers will be placed. There will also be a marked glare and even the lane lines may cause discomfort. Athletes should invest in good-quality all-round sunglasses for training. In some disciplines it may also be possible and more comfortable to wear them while competing.

Athletes may find it more comfortable to wear socks (cotton) inside their shoes. Not only will they absorb sweat, but they will also reduce the risk of blisters and the uncomfortable feeling of heat on the feet caused by a very hot track surface.

Athletes with spinal disabilities often have to remain in the open arena, for example, during throwing events. They should use an umbrella for shade; this can be fixed to the handle of a wheelchair that is used between throws.

ASTHMA

Asthma is common in athletes. Larsson et al. (1993) found that, out of 42 Swedish elite cross-country skiers, 24 had asthma and 31 had asthma-like symptoms. In 1984, 11.2 per cent of the athletes in the United States' Olympic Team had asthma induced by exercise (Voy, 1986). In 103 elite runners from Finland's adult and junior national teams, 16 had asthma documented by a physician, and all 16 had symptoms like exercise-induced asthma. Of the 87 remaining, 24 (28 per cent) had allergies, and of these, 14 (58 per cent) reported symptoms like exercise-induced asthma, as did 23 (37 per cent) of the 63 runners without asthma or allergies. Symptoms like exercise-induced asthma were reported in 41 (40 per cent) of the 103 runners during indoor training compared with 31 (30 per cent) during outdoor training (Tikkanen & Helenius, 1994).

Asthma and exercise-induced asthma are most common in people with known allergies. The pollen season is a time of increased incidence of asthma and hayfever. This is usually during summer months, but it is also possible for asthma to be triggered by aeroallergens in indoor training facilities in winter. Others may suffer only with particular pollutants. It is important for athletes to identify any trigger factors and try to avoid them if necessary. If an athlete has a known pollen sensitivity, it is important to identify the season in which the particular pollens occur. Before travelling abroad the athlete should check whether it is their bad pollen season in the destination country (*see* Dawood, 2002). It is possible that there may be different allergenic pollens in foreign countries. Animal dander is another notorious allergen and affected athletes should check ahead for the presence of animals, particularly if they plan to stay in domestic accommodation or in rural areas.

Breathing cold air (which always has a low relative humidity), and dry air itself, are known triggers for asthma, especially when there is hyperventilation, as in vigorous exercise (Lloyd, 1986). However, hypocapnoea (a low carbon dioxide pressure in the arterial blood) is induced by strenuous exercise, and this can also induce some resistance to airway flow (Newhouse et al., 1964), as can hypoxia (a low level of oxygen in the air) (Nadel & Widdicombe, 1962; Sterling, 1968). Under certain circumstances, these various factors may interact in the production of asthma-like symptoms (Pollard & Pollard, 1994).

Many pollutant gases (*see* Chapter 7) can affect the respiratory systems of non-asthmatic subjects, and asthma sufferers are even more likely to be affected by such gases. Most of the studies in North America were in summer, when the exposure was to ozone, acid aerosols and particulates (acid summer haze), whereas in the UK the studies have been mainly in winter, with particulates, nitrogen oxides and sulphur dioxide (Ayres, 1994). All athletes should take precautions to minimize exposure to these gases, and known asthmatics should take extra precautions against exposure to known air pollutants. Ozone is also known to enhance the asthmatic effect of an allergen (Ayres, 1994).

During the night of Friday 24 June 1994, there was a severe thunderstorm in London and this was accompanied by an epidemic of asthma attacks requiring emergency attendance at hospital: 640 compared with the normal expected 66 (Thames Regions et al., 1996). Of these patients, 294 gave a previous history of asthma and 403 had a history of allergy in the form of hayfever; some obviously had both. However, for 283 this was the first attack of asthma, although some may have had hayfever. Of those attending hospital, 104 were admitted, including 5 to an intensive care unit.

This epidemic was associated with a high level of grass pollen aeroallergens 9 hours earlier, and with a sudden increase in wind, and a fall in temperature 4 hours earlier. Celenza et al. (1996) identified the sudden fall in temperature and air pressure, as well as a pronounced increase in humidity, rainfall and lightning strikes. In a separate analysis of this and other outbreaks of thunderstorm-associated asthma epidemics around the world, the same authors found additional association with temperature inversions, fog, high levels of ozone and various organic dusts, including fungal spores and rye grass pollen. For non epidemic asthma attacks, the association was mainly with air pollution, but also with apparently contradictory meteorological factors, such as rainfall, high humidity, low wind speed, high air pressure, and high and low temperature.

A certain number of athletes may not be affected under normal conditions but may become symptomatic at particular times of year, when their trigger pollens are at a high level, or when they are exposed to animal dander. Others may suffer only with particular pollutants. It is important for athletes to identify any trigger factors and try to avoid them if necessary. Known asthmatics should take extra precautions against exposure to known air pollutants.

Medication routines should not be altered from the norm, unless warnings about air quality are given out. In this case, the dose should be doubled immediately and twice daily should be increased to three times daily (Ayres, 1994).

Any athlete competing at the top level should be aware of the IOC regulations governing the use of normal inhaled therapeutic agents (*see* IOC and

WADA websites). They must be declared before competition even if they are seldom used. The IOC and WADA require the athlete to undergo a test to prove that bronchospasm is induced by exercise and is reversed by beta agonists. There are several methods of producing bronchospasm but the eucapnic hyperventilation (EVH) challenge is the most sensitive in asymptomatic athletes (Dickinson et al., 2006). In eucapnic hyperventilation, the subject has to over-breathe as much as possible. This would normally mean that the CO_2 level in the blood would fall but the testers add CO_2 to the inspired air to ensure that the blood level of CO_2 remained normal.

Minimizing the effects of air pollution for the asthmatic

- Asthmatics should be particularly careful when high sulphur dioxide (SO_2) levels prevail.
- Asthmatics should avoid dry, cold air when possible.
- If possible, asthmatics should exercise in a warm, humid environment, such as a swimming pool. If swimming is not a desired exercise mode, it may be possible to run on a treadmill in the pool area and still take advantage of the pool environment.
- Wearing a mask or scarf will help warm cold air and reduce the inhalation of pollutants.
- Breathing through the nose will help warm and humidify the air.
- Athletes should take their medication prior to exercise and have their inhaler readily available during exercise (carrying it or entrusting it to a back-up team member).
- When warnings about air quality are given out, normal doses of medication should be doubled and, if being used twice daily, the frequency should be increased to three times daily.

DIABETES

While abroad, insulin-dependent diabetics are vulnerable to problems that can upset their control, and they must take extra care. Vomiting may occur due to motion sickness, food poisoning or exotic food. When travelling, everyone is exposed to more and different types of infection, and any infection upsets glucose control. Also, any wound is more liable to become infected. It is often difficult to maintain the same diabetic control as at home, because activity levels may change markedly, along with diet. (*See also* Dawood, 2002.)

Diabetes does not predispose to the development of altitude sickness (Moore et al., 2001). However, hypoglycaemia (low blood-sugar levels), which occurs more commonly during hill-walking than in most other activities, is difficult to recognize – headache and lightheadedness are also symptoms of altitude sickness, and paraesthesia (tingling in fingers and feet) is also associated with the use of acetazolamide (Diamox).

When anorexia (loss of appetite) occurs secondary to acute mountain sickness (AMS), diabetics tend to compensate for reduced carbohydrate intake by reducing insulin to try and prevent hypoglycaemia (low blood sugar) during exercise. This can lead to hyperglycaemia (high blood sugar) and ketosis (ketones in the blood, which are a sign of the body having to metabolize fat instead of carbohydrate). If AMS then develops, with nausea and vomiting, the glycaemic (blood sugar) status can deteriorate into frank ketoacidosis (when ketosis becomes severe the blood becomes very acidic and this is a dangerous state), and death. The tendency to dehydration at altitude will worsen the situation. The use of acetazolamide (Diamox) may contribute to acidosis (acidity of the blood). Since symptoms are unreliable warning signs for hypoglycaemia, diabetics have to rely more on results of blood-glucose testing. However, above 3,700m, blood glucose meters give readings 60–80 per cent lower than normal, especially if the ambient temperature is low. Some meters may even cease to function at very low temperatures.

Diabetics have to make their own decision about going to high altitude. They must be aware of the dangers, especially the risk of AMS with vomiting, and the fact that they are a long way from health-care facilities. Diabetics should ascend more slowly and diabetics with significant retinopathy (damage to their eyes) should not ascend to high altitude.

SICKLE-CELL DISEASE (*see also* page 185)

Sickle-cell disease is a condition found in people of African origin, particularly those from West and Central Africa. They have abnormal haemoglobin Hb5 (normal is HbA) – the red cells in the blood have a tendency to change from their normal round disc shape into sickle shapes. The sickled cells have a greater risk of causing blood clots, which may occur in any organ of the body. Those with Hb SS or Hb SC have the full disease, with a high incidence of anaemia and clotting crises in any organ. This may be accompanied by bone pain, abdominal pain, stroke and other neurological crises, and occasionally splenic rupture. Those with Hb AS have sickling trait – the cells are less likely to sickle, but there is an increased incidence of sudden death in sustained exhausting exercise. One of the precipitating factors is hypoxia, which means that those with the condition are likely to develop problems at altitude.

Treatment is oxygen, fluids and pain relief, and descent if at altitude.

Athletes of African origin or ancestry must be checked for sickle-cell disease and trait. In the USA, 13 per cent of people of African origin show a sickling trait. Those with sickle-cell disease or trait should not climb to high altitudes or go to altitude for athletic training.

MEDICINES AND MEDICAL CONDITIONS

The anti-malarial tablets Chloroquine and Mefloquine (Larium) will exacerbate epilepsy and Mefloquine should not be taken by those with a history of mental disorder.

It has been advised that those who are on Warfarin and oral anticoagulant medication or who are epileptic should not take melatonin (Herxheimer and Waterhouse, 2003) (*see also* page 131). In any case, athletes should consult their GP about the possible use of melatonin.

CHAPTER 10

Diseases and Illnesses

During a sporting career there are many diseases and illnesses that may be acquired, not always as a direct result of training or competition. The risk of disease, illness or injury is constantly present during the necessary activities that fill the gaps between training and competition. This chapter will focus on infections and methods of prevention (*see also* Chapter 8). In many cases even the possibility of these infections is unknown or ignored.

There are many routes for infections, and many different types that can be acquired, and they do not all occur abroad. The list of possible infections worldwide is so large that only the commoner ones, and the ones most likely to be encountered, will be included; for others, *see* Dawood (2002). Some infections can be prevented and these will be detailed. Diseases and illnesses are grouped in tables according to the methods of transmission.

DISEASES TRANSMITTED BY VARIOUS MEANS

Risks of transmission during sport are slight (Kordi & Wallace, 2004), but any player who is bleeding should be removed from play until the bleeding has stopped. Any blood spilled on the pitch should be cleaned up. In field hockey these points are mandatory and the artificial playing surface is cleaned with a special sterilizing agent before play restarts.

MANAGEMENT OF DIARRHOEA AND VOMITING

Most cases of diarrhoea and vomiting will settle in 48–72 hours without any treatment. The most important element in treatment is to ensure that adequate fluid is given, to prevent dehydration. In most situations it is perfectly possible to maintain fluid balance by mouth.

Antibiotics are of no value in the majority of cases, and should not be given unless a specific organism has been identified or is suspected (Dawood, 2002).

Antidiarrhoeal agents are rarely absolutely necessary, and they may even be dangerous if given early. They should never be given if there is fever or bloody diarrhoea. The use of these agents can be justified when sanitary arrangements are difficult, for example, for long journeys, or for important short-term events.

Table 9 Diseases transmitted by direct contact

Disease	*Source*	*Initial effect*	*Risk area*	*Special risk*	*Late effects*
Impetigo (bacteria)	Skin to skin. Via clothing or sharing towels	Rash on face	ww	Contact sports	
Erysipelas (bacteria)	Skin to skin. Via clothing or sharing towels	Painful red swollen face	ww	Contact sports	
Tinnea barbae (fungal)	Skin to skin. Via clothing or sharing towels	Rash on face	ww	Contact sports	
Herpes (virus)	Skin to skin, plus droplet spread	Rash on face	ww	Contact sports	Cold sores
Athlete's foot (fungus)	Walking on damp floors	Lesions between toes	ww	Showers and swimming pools	
Verucca (virus)	Walking on damp floors	Painful nodules	ww	Showers and swimming pools	

ww = worldwide

Table 10 Diseases transmitted by direct contact with blood

Disease	*Source*	*Risk area*	*Special risk*	*Late effects*
Hepatitis B (virus)	Infected blood plus broken skin in recipient	ww	Contact sports; sexual intercourse; contaminated needles and surgical instruments	Liver damage, death
HIV/AIDS (virus)	Infected blood plus broken skin in recipient	ww	Contact sports; sexual intercourse; contaminated needles and surgical instruments	Multisystem damage, death

ww = worldwide

Table 11 **Diseases transmitted by direct contact with water**

Disease	*Source*	*Initial effect*	*Risk area*	*Special risk*	*Late effects*
Vibrio vulnificus	Sea water	Wound infection	Gulf of Mexico		Septicaemia, death
Schistosomiasis	Fresh water, through intact skin	Fever, skin itch	ww		Multisystem
Leptospirosis (Weil's disease)	Fresh water; rodent urine through broken skin	Influenza-like	ww[1]	Canoeing, windsurfing, swimming	Kidneys
Conjunctivitis (eye), otitis externa (ear), folliculitis (skin)	Infections due to prolonged immersion		ww	Prolonged training in water	

ww = worldwide

Gastroenteritis, hepatitis A, pneumonia, salmonellosis, polio, shigellosis, meningitis, skin infections and eye, ear, nose and throat infections can all be acquired while swimming. For information on neurotoxicity from algal blooms, *see* Chapter 7 and Dawood (2002).

[1] The distribution is now worldwide, although it is salutary to think that sport has probably been the cause of this (Dawood, 2002). An expedition endurance event took place in Borneo in 2000 with 312 athletes from 26 countries. As many as 50 per cent of the participants are believed to have been infected with leptospirosis and taken the disease home with them.

Table 12 **Diseases transmitted by direct contact via droplets**

Disease	*Source*	*Initial effect*	*Risk area*	*Special risk*	*Late effects*
Tuberculosis (bacteria)	Droplets, infected milk	Lungs, intestines	ww	Crowds	Lungs, death
Diphtheria (bacteria)	Droplets, secretions, contact with skin lesions	Fever, sore throat	ww	Crowds	Throat membrane, death
Meningitis (bacteria virus)	Droplets	Fever, headache	ww	Crowds	Death
Influenza (virus)	Droplets	Fever, upper respiratory infection	ww	Crowds	
Common cold (virus)	Droplets	Breathing system	ww	Crowds	
Legionnaire's	Droplets	Breathing system	ww	Air conditioning, spas, showers etc.	Death

ww = worldwide

Table 13 **Diseases due to contaminated food or drink or failure in hygiene**

Disease	*Source*	*Initial effect*	*Risk area*	*Special risk*	*Late effects*
E. coli	Bacteria	Diarrhoea &/ or vomiting	ww	Contaminated fields or fresh-water lakes	
Salmonella	Bacteria	D &/or V	ww		
Typhoid	Bacteria	D &/or V	ww	Where food and water liable to faecal contamination	Death
Clostridium	Bacteria	D &/or V	ww		
Staphylococcus	Bacteria	D &/or V	ww		
Campylobacter	Bacteria	D &/or V	ww		
Bacillus cereus	Bacteria	D &/or V	ww		
Cholera	Vibrio	D &/or V	ww	Water, food, shellfish	Dehydration, death
Vibrio parahaemolytica	Bacteria	D &/or V	Japan, American coast	Shellfish	
Vibrio vulnificus	Bacteria	D &/or V, septicaemia, wound infection	Gulf of Mexico	Raw seafood, water into wounds	Death
Amoebic dysentery	Protozoan	D &/or V	T		Abscess liver, lung
Giardia	Protozoan	D &/or V	ww	Watersports	Chronic diarrhoea
Cryptosporidium	Protozoan	D &/or V	ww		
Toxoplasma	Cat faeces, cysts in undercooked meat	Fever	ww		Most tissues, including eyes
Hepatitis A	Virus	Varied	ww	[1]	Liver
Polio	Virus	Fever	ww		Meningitis, encephalitis, paralysis
Ciguatera	Fish: grouper, red snapper, barracuda, shellfish	Fever, gastro-enteritis	T	Reef fish	Death
Paralytic toxin	Shellfish	Slow heart, muscle paralysis	Carib		Death
	Shellfish	Fever	ww		
	Pesticides	Fever	ww		
	Drugs[2]	Fever			

ww = worldwide; T = all tropical areas

[1] Hepatitis A is spread by the faecal–oral route, frequently by person to person. Other sources are food handlers with raw or undercooked food, and raw or inadequately cooked shellfish grown in water contaminated by raw sewage. It is also possible to contract Hep A through swimming in coastal sea water. Viruses live longer than bacteria in sea water and swimmers may ingest viruses while swimming.

[2] Including Proguanil and Chloroquine.

Table 14 **Diseases transmitted by insect bites**

Disease	*Source*	*Risk of 'bites'*	*Risk area*	*Special risk*	*Late effects*
Malaria	Mosquito	Night, indoors or out; mainly rural	T		Recurrent fever, death
Yellow fever	Mosquito	Day, indoors or out; urban or rural	[1]		Bleeding into eyes, gut. Kidney, liver and heart failure; death
Dengue	Mosquito	Day, indoors or out; urban or rural	T		Haemorrhagic form
Jap Enceph	Mosquito	Evening and night, indoors or out; urban or rural	[2]	Paddy fields	Encephalitis; death
Other enceph	Mosquito	Evening and night, indoors or out; urban or rural	T		Encephalitis
Filariasis	Mosquito	Evening and night, indoors or out; urban or rural	T		Swelling of glands or leg; cough and wheeze
West Nile fever	Mosquito	Evening and night, indoors or out; urban or rural	[3]		Encephalitis
Sleeping sickness	Tsetse fly	Day, outdoors; rural	T, Af		Brain
Onchocerciasis	Black flies	Day, outdoors; rural	T	Fast-flowing water	Blindness
Leishmaniasis	Sand flies	Evening, indoors or out; urban or rural	[4]		Skin, erosion of mouth and nose (America)
Kala-azar visceral leish	Sand flies	Evening, indoors or out; urban or rural	[4]		Spleen, liver, bone marrow
Lyme disease	Tick (sheep, deer)	Day or night	[5]	Long grass, bracken	Multisystem
Encephalitis	Tick	Day or night	E&C Eur	Long grass, bracken	Encephalitis
Typhus	Tick	Day or night	T	Long grass, bracken	Severe illness
	Lice	Day or night	ww	On body hair or clothes	Death
	Mites	Day or night	As, Aus	From undergrowth or people	Fever
	Fleas	Day or night	ww	Markets, food, rubbish dumps	General illness
Chagas	Triatomic bug	At night, in beds	S Am		Heart damage
Plague	Fleas	Night or day, indoors or out	ww		High fever, muscle pain; swelling of glands (bubonic); lungs (pneumonic); septicaemia; death
Irritation	Midges	Evening, outdoors; rural	ww		
Irritation	Bed bugs	At night, in beds	ww		

ww = worldwide; T = all tropical areas; Af = Africa; Am = America; As = Asia; Aus = Australia; Eur = Europe; N or S or C or W or E = North or South or Central or Eastern or Western

[1] West and Central Africa and South and Central America. [2] Japan, China, India, Southeast Asia, Indonesia, now reached Northern Australia. [3] Africa, Southern Europe, Middle East, part of Far East, now widespread in USA. [4] North and Central America, Mediterranean basin, parts of Africa, the Middle East and Asia. [5] North America, Northern Europe and Asia, Coastal Australia.

If bitten by a tick, it is essential to remove the tick as soon as possible. To cause Lyme disease the tick normally needs to be attached for at least 12 hours. To remove a tick, grasp as far forward as possible and turn anticlockwise, taking care not to break off the head. Butter, paraffin or burning with a cigarette increase the risk of Lyme disease because the tick regurgitates stomach contents. A single dose of 200mg Doxycycline given within 72 hours of a tick being removed can prevent the development of Lyme disease (Nadelman et al., 2001)

A simple dipstick test has been developed to provide a cheap and accurate 15-minute test for plague, to allow early treatment (Chanteau et al., 2003).

VACCINATIONS AND IMMUNIZATIONS

There are only a few diseases that can be prevented by immunization and the situation should be assessed well before the planned travel. The requirements for each country can be obtained from a General Practice travel clinic or from the relevant websites (*see* page 169). Some immunizations will be noted as recommended while others are 'sometimes' recommended. The decision about which immunizations to take depends on a number of factors, including length of time in the country, and expected contact with local people.

Table 15 **Diseases for which vaccinations are available**

Disease	*Recommendations*
Cholera	No, not for normal travel and activities
Diphtheria	Yes, for everybody
Hepatitis A	*See* below
Hepatitis B	Yes, if likely to have contact with blood
Japanese encephalitis	Yes, if spending one month or more in an endemic area
Meningococcal meningitis A & C	No, only advisable in high-risk areas for teachers, health workers, backpackers and residents. Also for pilgrims to Mecca during Hajj
Polio	Yes, for everybody
Rabies	No, only for long periods, remote areas, or vets in high-risk areas. *See* below
Tetanus	Yes, for everybody (provides cover for 10 years and is renewable)
Tick-borne encephalitis	No, unless spending prolonged periods in forests or woodlands in Europe
Tuberculosis (TB)	Yes, especially if living in local conditions in developing countries
Typhoid	No, except for India and those living under local conditions in developing countries
Yellow Fever	*See* below

As hepatitis A is a food- or water-borne infection, all sports people would be well advised to arrange immunization before departure. Two doses of vaccine 2–4 weeks apart provide immunity for 1 year and a further booster after 6–12 months provides protection for 10 years.

Yellow fever has special requirements. For some countries a certificate of vaccination is compulsory for entry, while for other counties this requirement extends only to those who have come from a yellow fever endemic area. Any sportsman should have inoculation against yellow fever if they are entering an endemic country. Since inoculation provides cover for 10 years but only becomes effective 10 days after the inoculation, those who may have to travel to places at short notice might consider getting it done as an advance precaution. It can be renewed after 10 years. Yellow fever inoculations can only be given in specially registered centres.

For information on all vaccinations and immunizations, every individual should consult their GP or local travel clinic, as well as the relevant websites (*see* page 169) and Dawood (2002).

MALARIA

After the mosquito bite, the parasite develops in the liver before reappearing in the bloodstream to cause the attack of malaria. There are four strains of the malaria parasite: Plasmodium (P.) falciparum, P. vivax, P. ovale and P. malariae. Of these, P. falciparum is the most dangerous, and may rapidly kill the subject.

There is, as yet, no vaccination against malaria, although work is in progress (Dobson, 1996) and appears promising (Alonso et al., 2005). At present the official recommendation is to take tablets prophylactically. However, there are a number of problems with this system:

- The medication recommended varies between countries, depending on the drug resistance of the local parasites. This resistance changes over time and the information has to be kept up to date.
- The drugs have to be started some time before arriving in the country (the exact time depends on the drug used), and continue throughout the stay and for 4 weeks after returning home (because the onset of malaria may be delayed).
- All the preventive drugs used have side effects, which vary according to the drugs and to the susceptibility of the individual. Some of the side effects can be severe – for example, diarrhoea and vomiting with the Proguanil/Chloroquine combination, neuropsychiatric side effects in 0.1–1 per cent of those taking Mefloquine (Larium), and sun hypersensitization with Doxycycline. These side effects may be enough to make people stop the drugs, or avoid taking them in the first place. They are relatively rare, and some people suffer no side effects at all. In general, the risk of side effects is greatly outweighed by the risk of death with P. falciparum infection.
- No prophylaxis is 100 per cent effective and it is still possible to contract malaria even when the drugs have been taken.
- Chloroquine may exacerbate epilepsy and photosensitive psoriasis.
- Mefloquine should not be taken by those with epilepsy or a history of psychiatric disorder.

In the highly competitive world of modern sport, even the slightest factor may mean the difference between winning and losing. Because of this, many athletes may opt not to take the anti-malarial tablets, but this will leave them vulnerable to malaria and an infection with P. falciparum, which may kill quickly.

One option is not to take the antimalarial tablets but, in case of a high fever, testing the blood for malaria, and then starting treatment if the result is positive. Traditionally, this has not been practical since the diagnosis depended on a sample of blood being sent to a laboratory and a slide being examined. In many parts of the world there are no laboratories with the technical equipment or the skills required to test the sample. Even in the UK, sending the blood to a laboratory meant a wait of a few days before a result was available.

Today, it is possible to use one of the recently developed dipstick tests for detecting malaria from a drop of blood. The result is available in a few minutes and the tester needs no training. The most recent test, Optimal IT (*see* page 169), can identify the three less dangerous forms of malaria in the absence of P. falciparum. It can also identify P. falciparum, but the result obscures any presence or absence of the other three. It has a high degree of sensitivity and specificity and can be used to monitor the effect of treatment (Moody and Chiodini, 2002).

OTHER INFECTIONS

Brucellosis

Sufferers with brucellosis ofen present with unexplained lethargy and fatigue which is often attributed to psychological causes. It has been suggested that the cause of Florence Nightingale's prolonged illness was brucellosis (Young, 1995).

Rabies

A number of animals can carry the rabies virus, including bats. However, in practice, the dog is the main source of rabies in humans. Transmission can occur

Table 16 **Other infections**

Infection	*Source*	*Initial symptoms*	*Risk area*	*Late effects*
Brucellosis	Unpasteurized milk products including goat's cheese	Non-specific	ww	Prolonged disability
Listeriosis	Unpasteurized milk products including goat's cheese	Non-specific	ww	Disability
Rabies	Animal bite, usually dog	None	ww	Death

ww = worldwide

Table 17 **Worm infestations**

Worm	*Route of entry*	*Initial symptoms*	*Risk area*	*Late outcome*
Round	Eggs contaminate food	Wheeze, rashes	T	Gut, may cause obstruction, or jaundice
Whip	Eggs contaminate food	None	T	Gut, abdominal pain, dysentery
Thread	Eggs on food or dust	Anal itch	ww	Gut, minor symptoms
Herring	Larval worm in fish, raw or pickled sea fish or squid, sushi	Abdominal pain and vomiting	Japan	Gut
Tape (beef)	Larval cysts in meat; undercooked beef	None	ww	Gut, colic
Tape (pork)	Larval cysts in pork; undercooked pork	None	ww	Gut, colic; *see* cysticercosis
Cysticercosis	Eggs of pork tapeworm; poor hygiene	None	Lat Am, Af, India	Cysts in brain, heart, muscle
Hydatid	Eggs of dog tape-worm contaminate food and fingers	None	Mid East, Lat Am, N&E Af	Cysts in liver, lung, brain, anaphylaxis if cyst bursts
Trichinosis	Larval worms in undercooked pig or horse meat	Diarrhoea	ww	Muscle pain, weakness, swelling of face and eyelids
Toxicara	Eggs of dog round-worm contaminate food and fingers; in sand, soil and playing fields[1]		ww	Liver, lung, eye, brain
Hookworm	Soil contaminated with human faeces; through skin of bare feet or in long wet grass	Cough, wheeze, upper abdominal pain	Warm temp and T	Anaemia
Strongyloidiasis	Soil contaminated with human faeces; through skin of bare feet or in long wet grass	Cough, wheeze, upper abdominal pain, plus skin tracks	Warm temp and T	Self-infecting, malabsorption, diarrhoea
Larva migrans	Direct contact of any part of body with contaminated soil or sand	Skin tracks	T	Abdominal pain and vomit

ww = worldwide; T = all tropical areas; Af = Africa; Am = America; As = Asia; Aus = Australia; Eur = Europe; N or S or C or W or E = North or South or Central or Eastern or Western; Warm temp = warm temperature zones

[1] Toxicariasis is more likely to cause problems in children than adults (Gillespie, 2001)

following a nip (Solomon et al., 2005) or even a lick over broken skin (Dawood, 2002). The danger is that, because the incubation period is from 1 to 3 months, by the time the person knows for certain that infection has occurred it is too late. There are several essential steps. Pre-exposure vaccination is available and should be used if the person is going to be at high risk, or if he is particularly apprehensive. However, pre-exposure vaccination does not eliminate the need for treatment after rabies exposure (Pounder, 2005). It is important to notify medical services immediately after any bite, and to start treatment. The wound must be scrubbed with antiseptic (under local anaesthetic), and immunization started within 24 hours (Dawood, 2002; Pounder, 2005).

Tetanus and Gas Gangrene (Clostriduim Welchii)

These organisms thrive in contaminated wounds where there is lack of oxygen, such as deep penetrating wounds, or if there is a lot of tissue damage.

AFTER RETURNING HOME

- If the traveller has to consult the doctor, he should always inform the doctor that he has been abroad, and where and when.
- Anti-malaria medication must be continued for the full 4 weeks.
- If there are no symptoms, a routine examination is unnecessary, unless the person has had illness abroad or has been living in a risk area for a long time.
- Some infections may cause problems on return, including salmonellosis, campylobacter enteritis, shigella dysentery, malaria, giardiasis, viral hepatitis, enteric fever, and cholera. Intestinal helminths (worms) may be present after return, but only if the person has been in an area of specific risk.
- Fever or influenza-like symptoms must be investigated. Malaria is a possibility up to 3 years after return, even if the tablets were taken. Falciparum malaria usually shows within 3 months. Other possibilities include dengue, hepatitis, pneumonia (possibly legionnaire's disease), typhoid, tick typhus, acute schistosomiasis. Pulmonary embolism and drug fever are other possible causes, and there may also be a common non-exotic cause.
- Jaundice may be due to a hepatitis virus or malaria.
- Most cases of diarrhoea require only symptomatic treatment. If fever is also present, consider bacillary dysentery and malaria. If there is blood in the stools in an acute infection, consider bacillary or amoebic dysentery, strongyloides stercoralis infestation, chronic schistosomiasis (also blood in urine). If the diarrhoea is chronic, giardia and amoebic infection are possibilities.
- Respiratory symptoms are common, but need to be investigated.
- A serpentine red track on the skin that is itchy and blistering may be due to the larvae of the dog tapeworm. A spreading rash on the skin may indicate Lyme disease.
- Vague symptoms may be due to viral hepatitis or brucellosis. Chronic fatigue may occur in 2 per cent of those who have had cigatura poisoning (Pearn, 1997).

APPENDIX I

Some Aspects of Exercise Physiology

Knowledge of human physiology will help the athlete understand and appreciate some of the changes that occur when humans exercise, and the ways in which exercise may be affected by different environments.

CIRCULATION OF BLOOD

Blood returning from the cells and muscles is carried in veins (the venous blood), with a reduced oxygen (O_2) and increased carbon dioxide (CO_2) content. It enters the right side of the heart and is pumped (via pulmonary arteries) to the lungs where O_2 is taken up and some CO_2 is removed. The blood proceeds from the lungs, via the pulmonary veins to the left side of the heart, where it is pumped around the body in arteries (arterial blood), which branch into capillaries (small blood vessels). (*See* Figure 23.)

OXYGEN AND CARBON DIOXIDE

O_2 is necessary for life and for the maintenance of prolonged exercise. O_2 is present in the atmosphere and has a percentage concentration of 20.93 per cent. O_2 needs to be transferred from the atmosphere to the cells and muscles of the body, and it enters the body via the mouth and nose. The respiratory system (Figure 24) comprises the trachea, its branches, the bronchi (the tubes that conduct the air to the alveoli) and the alveoli (sacs with thin membranes that allow gas transfer between the lungs and the blood) in the lungs. Gas transfer takes place between millions of capillaries and alveoli. O_2 is transferred to the blood and is then carried to the cells, where it is used up, and CO_2 is produced.

The pressure of a given gas is directly related to the percentage concentration of the gas. This is known as the partial pressure and is shown as Pgas. The partial pressure denotes the pressure of a given gas in a mixture of gases e.g. the partial pressure of O_2 (PO_2) or CO_2 (PCO_2) in the blood. At sea level, barometric pressure in the atmosphere is 760mmHg. The PO_2 is 159mmHg (20.93per cent of 760mmHg) at sea level, but as the air coming into the lungs from the atmosphere mixes with air in the lungs that has already been used, the proportion of O_2 in the alveolar air falls and the resultant PO_2 in the alveoli is around 105mmHg. The venous blood (deoxygenated blood) enters the lungs via the pulmonary capillaries, and there is a transfer of O_2 from the air in the lungs to the blood in the pulmonary capillaries via diffusion – there is movement of O_2 from an area of high PO_2 to a low PO_2. As the diffusion takes

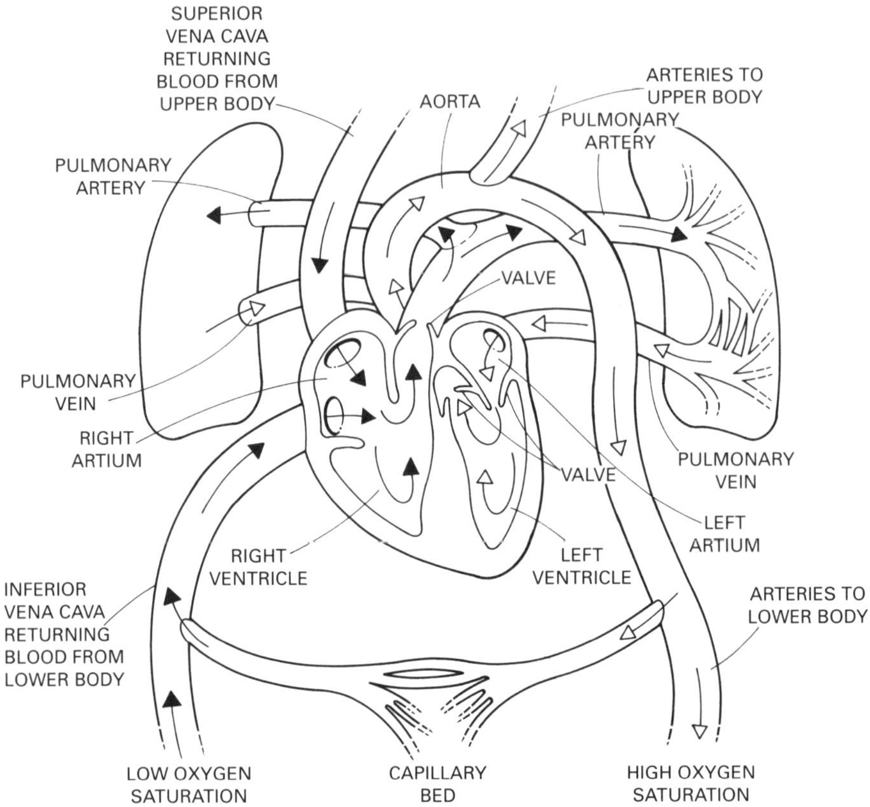

Fig 23. Circulation of blood in some areas of the human body.

place, the PO_2 rises in the pulmonary capillaries and becomes the same as the PO_2 in the alveolar air. The large gradient at rest at sea level between the PO_2 in the alveoli (105mmHg) and the PO_2 in the pulmonary capillaries (40mmHg) results in a rapid transfer of O_2 to the blood in the pulmonary capillaries. (This diffusion gradient – the difference between the venous PO_2 and alveolar PO_2 – is much less at altitude, resulting in limited diffusion.) Diffusion rates are faster when the difference in PO_2 is large compared with small differences in PO_2. As airflow and blood flow are not precisely matched within all areas of the lung, there is a decrease in the PO_2 of the arterial blood (arterial blood has a PO_2 100mmHg). (*See* Box below on partial pressure of oxygen (PO_2) and partial pressure of carbon dioxide (PCO_2).)

Partial pressure of O_2 (PO_2)

The partial pressure for O_2 (the PO_2) in the atmosphere is 159mmHg. As the O_2 passes through the body system, the PO_2 falls. The O_2 is used in the cells, so, by the time the blood returns through the veins, the PO_2 is much lower.

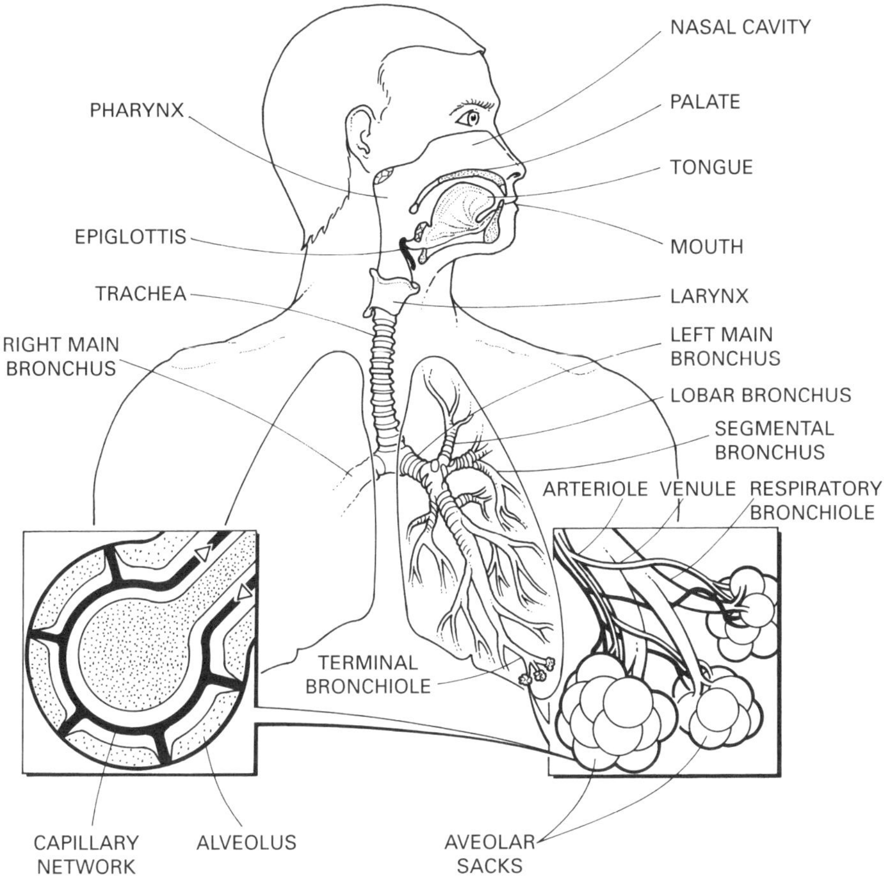

Fig 24. Main components of the respiratory system.

Summary of sea level partial pressure of oxygen (PO_2) and partial pressure of carbon dioxide (PCO_2) in the atmosphere, alveoli, arterial blood, cells and venous blood at rest

	Partial pressure of oxygen (PO_2)	*Partial pressure of carbon dioxide (PCO_2)*
Atmosphere	159mmHg	0.3mmHg
Alveoli	105mmHg	40mmHg
Arterial blood	100mmHg	40mmHg
Cells	40mmHg	46mmHg
Venous blood	40mmHg	46mmHg

Arterial and venous blood refer to values in systemic circulation

Oxygenated blood is carried from the lungs by the pulmonary veins to the left side of the heart, where it is pumped via the arteries and capillaries to the cells. The arterial blood enters capillaries with a PO_2 of 100mmHg while the cells have a PO_2 of about 40mmHg at rest (lower in exercise). The O_2 diffuses from the capillaries to the cells whereas CO_2 diffuses in the other direction from the cells to capillaries. CO_2 is produced in the cells, which have a PCO_2 of about 46mmHg at rest. The blood (venous blood) that returns to the right side of the heart (and then to the lungs) from the cells has a PO_2 of around 40mmHg and a PCO_2 of about 46mmHg. As the PCO_2 in the alveoli is 40mmHg, CO_2 diffuses from the high PCO_2 of 46mmHg in the pulmonary capillary to the alveoli.

CARDIAC OUTPUT (CO)

The CO is the volume of blood pumped by the heart in a minute. It is the product of heart rate (the number of beats per minute) and the stroke volume (SV) (the volume of blood ejected by each beat). At rest, the CO is about 5–6ltr per minute. For example, a SV of 75ml per beat and a heart rate of 70 beats per minute results in a CO of 5.25ltr per minute. At maximum, the CO of an untrained 20-year-old is about 20ltr per minute, whereas an elite endurance athlete may have a maximum CO of 38ltr per minute (maximum heart rate of 200 beats per minute and a stroke volume of 190ml per beat). A large stroke volume is a characteristic of elite endurance athletes. The larger the CO is during exercise, the greater the volume of O_2 that is carried to the muscles.

BLOOD

Blood is a fluid that contains several parts, including plasma (the watery component of blood) and red cells. There are billions of red cells in the body. The red cells have a protein called haemoglobin (Hb), which contains iron and binds O_2. Haemoglobin carries almost all of the O_2 in the blood. Production of red cells is controlled by the hormone erythropoietin, which is secreted mainly by the kidneys. Erythropoietin stimulates the bone marrow to produce red cells. A decrease in the delivery of oxygen to the kidneys will promote an increase in erythropoietin production. The haemoglobin concentration for males is around 15g per 100ml of blood and for females around 13.5g per 100ml of is blood. The principal factor that determines how much O_2 is carried by haemoglobin is the PO_2. Figure 25 shows the oxygen-haemoglobin dissociation curve (the pressures at which haemoglobin holds on to or gives up O_2). At sea level the PO_2 of the arterial blood is 100mmHg (arterial PO_2), and at that PO_2 the haemoglobin is 97 per cent saturated in the arterial blood. Note that if the PO_2 falls to 80mmHg there is very little decrease in haemoglobin saturation. At a PO_2 of 40mmHg the haemoglobin saturation is 75 per cent. When fully saturated, each gram of haemoglobin carries 1.34ml of O_2. Bearing the above in mind, it can be seen that the 'average' male carries O_2 at rest at sea level as follows:

> Arterial blood – 15g Hb (1.34ml O_2/g Hb) = 20.1ml O_2 per 100ml of blood (fully saturated)

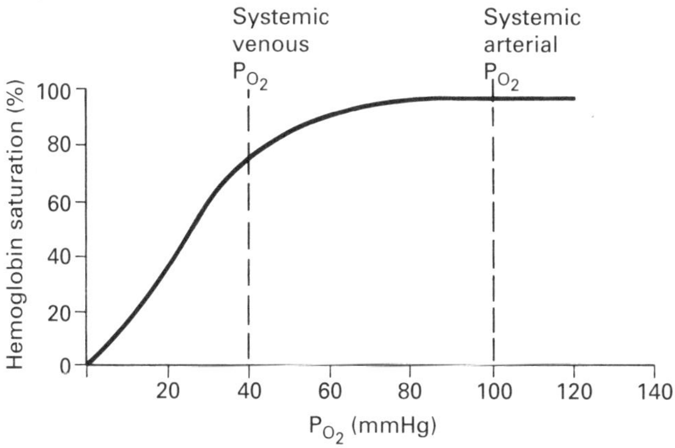

Fig 25. The oxygen-haemoglobin disassociation curve. This curve applies to blood at 37°C and a normal arterial hydrogen-ion concentration.

(From Vander, A., Sherman, J. and Luciano, D., *Human Physiology: The mechanisms of body function*, Sixth Edition, 1994. With permission from McGraw-Hill Companies)

Arterial blood 97 per cent saturated = 19.5ml O_2 per 100ml of blood

Venous blood 75 per cent saturated = 15.1ml O_2 per 100ml of blood

Arterial–venous O_2 difference (A–VO_2 diff; the difference between the O_2 content of the arterial and venous blood) = 4.4ml O_2 per 100ml of blood

The resting arterial PO_2 at sea level is 100mmHg

At a PO_2 of 100mmHg, haemoglobin is 97 per cent saturated

The resting PO_2 of the venous blood is 40mmHg

At a PO_2 of 40mmHg, haemoglobin is 75 per cent saturated

The plasma carries 0.3ml of O_2 per 100ml of blood at PO_2 of 100mmHg at sea level.

If haemoglobin is increased from 15 to 18g per 100ml blood, the arterial O_2 content of the blood is increased. Comparison of the arterial oxygen content for haemoglobin concentrations of 15g haemoglobin per 100ml of blood and 18g haemoglobin per 100ml of blood is given below.

The following example is for arterial blood fully saturated, and arterial blood at an altitude of 4,800m (the summit of Mont Blanc) when it is 78 per cent saturated:

Blood – 15g Hb (1.34ml O_2/g Hb) carries 20.1ml O_2 per 100ml of blood (fully saturated).

Arterial blood (15g Hb) – 78 per cent saturated carries 15.7ml O_2 per 100ml of blood.

Arterial blood – 18g Hb (1.34ml O_2/g Hb) carries 24.1ml O_2 per 100ml of blood (fully saturated).

Arterial blood (18g Hb) 78 per cent saturated carries 18.8ml O_2 per 100ml of blood.

Note that the arterial O_2 content is much higher with a haemoglobin concentration of 18g per 100ml of blood.

Haemoglobin

Haemoglobin carries almost all of the O_2 in the blood. The haemoglobin concentration for males is around 15g per 100ml of blood, and for females around 13.5g per 100ml of blood. The principal factor that determines how much O_2 is carried by haemoglobin is the partial pressure of oxygen (PO_2) (*see* page 157).

BREATHING

Minute Ventilation (VE)

The amount of air that is expired in a minute is called the minute ventilation (VE). This consists of the frequency of breathing (respiratory rate, or RR), and the tidal volume (TV). The TV is the volume of air that is moved during an inspiration or expiration. At rest, the minute ventilation (VE) is between 5 and 15ltr per minute. A resting VE of 6ltr per minute could be made up of a respiratory rate (RR) of 10 breaths per minute and a tidal volume (TV) of 0.6ltr.

Large, well-trained male athletes (for example rowers) may have a maximum VE of over 200ltr/min, with respiratory rates of about 60 breaths per minute and a tidal volume of about 3.5ltr (*see* Table 18).

Control of Breathing

The control of breathing at rest is normally via chemoreceptors in the brain (central chemoreceptors) that detect changes in acidity (H^+) in the brain's extracellular fluid resulting from differences in arterial CO_2. An increase in CO_2 increases VE, while a reduction in CO_2 lowers VE. Small increases in CO_2 can increase VE by 100 per cent. Knowledge of this has prompted swimmers who want to dive under water for some time, without breathing apparatus, to hyperventilate or over-breathe (blow in and out hard). This blows off (or reduces) CO_2 before submersion, and consequently reduces the diver's urge to breathe. A diver who 'blows off' CO_2 before an attempt to swim 50m underwater in the swimming pool can stay under water longer than one who does not blow off of CO_2. This strategy may result in loss of consciousness, and death, as the urge to breathe diminishes and O_2 levels may become so low that the swimmer loses consciousness.

Table 18 **Respiratory rate, tidal volume and minute ventilation (volume breathed out) (VE) at rest and maximum exercise for a large fit rower**

	Respiratory rate (breaths/min)	*Tidal volume (litres/breath)*	*VE (litres/min)*
Rest	10	0.6	6.0
Maximum	60	3.5	210

O_2 can also influence breathing. Although the percentage of O_2 (20.93 per cent) in the air at altitude is the same as at sea level, the atmospheric pressure is reduced at altitude. For example, at 3,700m (an altitude that promotes an increase in VE), the total atmospheric pressure is only about 489mmHg compared with 760mmHg at sea level. Therefore the PO_2 is reduced at altitude. The lower PO_2 (102mmHg) at 3,700m, compared with sea level (159mmHg), results in a lower PO_2 in the alveoli and arterial blood. The reduced PO_2 in the arterial blood is detected by peripheral chemoreptors (O_2 sensors in the arteries in the neck called the carotid bodies), and they signal the respiratory control centre in the brain to increase VE in an attempt to counter the effects of a lowered PO_2. The magnitude of VE increase at altitude can vary greatly between individuals.

A number of factors can contribute to the increase in VE during exercise but, despite many decades of research, the stimuli are not yet fully understood. The fact that VE increases very rapidly at the start of exercise and falls quickly at the end of exercise is suggestive of regulatory influences other than changes in arterial PCO_2 and H^+ concentration. Increases in core temperature and input from receptors in joints and muscles are two possible non-chemical influences on VE.

During moderate exercise there is little change in the level of CO_2 in the arterial blood – in high-intensity exercise, it actually falls – so it is not possible to attribute changes in VE during exercise to increases in levels of CO_2 in the arterial blood. In high-intensity exercise there is an increase in the acidity (H^+)

Oxygen influencing breathing

The partial pressure of oxygen (PO_2) is reduced at altitude. At an altitude of 3,700m (an altitude that promotes an increase in VE), the total atmospheric pressure is only about 489mmHg compared with 760mmHg at sea level. Therefore the PO_2 is reduced at altitude. The lower PO_2 at 3,700m results in a lower PO_2 in the alveoli and arterial blood.

The reduced PO_2 in the arterial blood is detected by the carotid bodies which signal to the respiratory control centre in the brain to increase VE in an attempt to counter the effects of a lowered PO_2.

of the arterial blood (associated with increases in lactic acid). The increase in H^+ concentration is detected by the peripheral chemoreceptors and this stimulates an increase in VE.

MAXIMAL OXYGEN UPTAKE (VO_2MAX)

Maximal oxygen uptake (VO_2max) is defined as the greatest amount of O_2 that can be used by the body per minute during exercise. Scores are given in litres per minute (L/min) or millilitres per kilogram of body weight per minute (ml/kg/min) when related to body weight. Typical scores for sedentary 20-year-old males and females are 45ml/kg/min and 40ml/kg/min respectively. Elite endurance male and female athletes have scores of about 80ml/kg/min and 70ml/kg/min respectively.

VO_2max is a measure of cardiorespiratory fitness. While a high score is a prerequisite for elite endurance performance, other factors are also predictive of endurance performance, including exercise economy and the lactate threshold. The intensity of exercise is often prescribed or measured using a percentage of VO_2max. For example, 90 per cent of VO_2max is deemed to be high intensity, 40–60 per cent VO_2max is moderate, whereas 30 per cent VO_2max is considered to be low intensity.

OXYGEN COST OF EXERCISE

In order to sustain exercise for long periods of time, the exerciser must be able to supply O_2 to the muscles, which must be able to take up O_2. The faster the speed (whether running, cycling or canoeing), the greater the O_2 cost for a given activity. A number of factors influence the O_2 cost, including running style, which means that some runners use less O_2 than others at a given speed. A runner with a lower O_2 cost is said to be more 'economic' than a runner with higher O_2 costs. It is advantageous to have a low O_2 cost for a given speed as this reduces the physiological strain on the runner, and that runner will be able to sustain higher speeds than a runner who has all the same physiological

Table 19 **Oxygen costs (ml/kg/min) for given treadmill running speeds and gradients**

Speed (mph)	*Flat surface*	*5 per cent incline*	*10 per cent incline*	*15 per cent incline*
5	30.3	36.3	42.4	48.4
8	46.4	56.0	65.7	72.5
10	57.1	69.2	81.2	93.3
15	83.9			

Data calculated from equations from the American College of Sports Medicine, Guidelines for graded exercise testing and Exercise Prescription, 1986, Lea and Febiger.

characteristics, but poor economy. Measurement of the O_2 cost of exercise can be useful in a number of areas. For example, it has been shown that the O_2 cost of walking on snow is greater than that of walking on clear ground. Shivering has also been shown to increase the O_2 cost of an activity.

Note:

- The O_2 costs refer to a steady speed sustained for a long period of time.
- The faster the speed, the greater the O_2 cost.
- Increasing the gradient results in a higher O_2 cost.
- Some speeds/gradients are not sustainable for individuals who have a VO_2max below the oxygen cost for a given speed/gradient. For example, a subject with a VO_2max of 40ml/kg/min cannot sustain a speed of 10mph, which has an O_2 cost of 57.1ml/kg/min.
- The speeds and gradients that result in very high O_2 costs cannot be sustained for very long, but an anaerobic contribution allows the athlete to run at very high speeds and gradients for a relatively short period of time.
- VO_2 for a given running speed has been found to be low in some world class marathon runners, including Frank Shorter and Derek Clayton, even compared with Jim Ryun, a former world record-holder for the mile.
- VO_2 is much higher for a given running speed for children than for adults.

APPENDIX II

Physiological Responses to Altitude

Description in detail with physiological terms

BAROMETRIC PRESSURE

Barometric pressure decreases as altitude increases and, though the percentage of oxygen (O_2) remains the same, the partial pressure of O_2 (PO_2) (*see* page 153) is lowered. The barometric pressure is 760mmHg at sea level and decreases with increasing altitude. For example, a barometric pressure of 543mmHg is found at 2,800m and 429mmHg at 4,700m (*see* Table 20, page 163). As the barometric pressure decreases, the inspired PO_2 (PIO_2) decreases in direct proportion to the fall in barometric pressure. This lowering of the PO_2 at altitude has an effect on the O_2-carrying capacity of the arterial blood. With increasing altitude, the PO_2 in the alveolar air (PAO_2) and, therefore, the PO_2 in the arterial blood (PaO_2) also falls. Table 20 gives examples of how the PAO_2 and the PaO_2 decrease with increasing altitude and decreasing PIO_2. At sea level PAO_2 and the PaO_2 are 101mmHg and 89mmHg respectively but at 4,700m they are 48mmHg and 45mmHg respectively.

Most of the O_2 that is carried by the blood is bound to haemoglobin (Hb). The amount of oxygen that is carried by haemoglobin is determined predominantly by the PO_2. A PO_2 of 100mmHg (arterial PO_2 at rest at sea level) results in an arterial oxygen saturation (SaO_2) of 97 per cent. As the PO_2 decreases, the SaO_2 falls. The SaO_2 does not decrease steeply until around a PO_2 of 60mmHg. Thus, the plateau in the oxygen-haemoglobin disassociation curve (*see* Figure 25, page 157) provides a 'reserve'. It also means that, for low mountains such as Ben Nevis, at 1,344m, there will be little difference in the SaO_2 at rest from the value at sea level. However, when at 3,660m the PaO_2 falls to 48mmHg, the SaO_2 is 85 per cent (*see* Table 20).

The amount of oxygen bound to haemoglobin in the blood has an important influence on the arterial oxygen content of the blood. The following demonstrates that the arterial oxygen content is lower at 3,660m (arterial blood – 85 per cent saturated) than at sea level (arterial blood – 97 per cent saturated):

Arterial blood – 15g[†] Hb* 1.34ml O_2 = 20.1ml O_2 per 100ml of blood (fully saturated)

Arterial blood 97 per cent saturated (sea level) = 19.5ml O_2 per 100ml of blood

Arterial blood 85 per cent saturated (3,660m) = 17.1ml O_2 per 100ml of blood

* 1g of haemoglobin can carry 1.34ml O_2 when it is fully saturated
† 15g of haemoglobin per 100ml of blood is a typical value for males at sea level.

It should be noted that the above values refer to rest. With exercise, even light exercise at 4,000m, the SaO_2 is likely to decrease from the resting values.

Table 20 highlights the following:

- barometric pressure decreases with increasing altitude;
- PIO_2, and also PAO_2 and PaO_2 decrease with increasing altitude;
- SaO_2 at 2,810m is 91 per cent. This is appreciably lower than sea level values;
- SaO_2 at 4,700m is 80 per cent. This is even more markedly lower than sea-level values;
- the alveolar PCO_2 ($PACO_2$) (mmHg) is much lower at altitude.

There are a number of consequences and adjustments to the body systems as a result of the reduced PIO_2. There is an increase in the volume of air breathed out per minute (VE) (*see* page 159) at altitude. The low PO_2 in the arterial blood is detected by the carotid bodies, which communicate with the respiratory control centre in the brain, and VE is increased. This VE increase is advantageous for the individual at altitude as it increases PAO_2, which would not be the case if there were no increase in VE. There is marked individual vari-

Table 20 **Barometric pressures, oxygen pressures, arterial oxygen saturation and alveolar PCO_2 at various altitudes**

Altitude (m)		*0*	*2,810*	*3,660*	*4,700*
Pressures					
BP	(mmHg)	760	543	489	429
PIO_2	(mmHg)	159	114	102	90
PAO_2	(mmHg)	101	62	58	48
PaO_2	(mmHg)	89	60	48	45
SaO_2	(per cent)	97	91	85	80
$PACO_2$	(mmHg)	41	34	30	27

BP = barometric pressure
PIO_2 = inspired PO_2
PAO_2 = alveolar PO_2
PaO_2 = arterial PO_2
SaO_2 = oxygen saturation in the arterial blood
$PACO_2$ = alveolar PCO_2

Adapted from McFarland, R. A. and Dill, D. B. (1938), A comparative study of the effects of reduced oxygen pressure on man during acclimatization, *Journal of Aviation Medicine*, *9*, 18–34. (Used with permission from the Aerospace Medical Association, 320 S. Henry Street, Alexandria, VA 22314. USA.)

ability in the response to the reduced PO_2.

COMPARISON OF RESPONSES AT SEA LEVEL AND AT 4,000M

Stenberg et al. (1966) compared the physiological responses at sea level and acute exposure to 4,000m using a cycle ergometer in a laboratory. At sea level, the group's average VE at 150 watts was 59.0L/min but, when asked to exercise at the same intensity at 4,000m VE rose to 93.7L/min. During submaximal exercise at 150 watts, the SaO_2 was 95 per cent at sea level, and 69 per cent at 4,000m. At sea level, maximal oxygen uptake (VO_2max) (*see* page 160) was 3.46L/min. This was reduced to 2.50L/min at 4,000m.

There are a number of points to emphasize:

- in submaximal exercise (150 watts) the VE is much higher at 4,000m compared with at sea level;
- VO_2max is lower at 4,000m compared with at sea level;
- the SaO_2 at 4,000m is lower than at sea level;
- the per cent VO_2max at 150 watts at sea level is 63 per cent;
- the per cent VO_2max at 150 watts at 4,000m is 89 per cent;
- thus, the relative exercise intensity is higher at 4,000m – the subject is operating at a higher percentage of VO_2max at altitude than he would have done at sea level.

The increased VE results in the 'blowing off' of CO_2 and an increase in the pH (reduced acidity, increased alkalosis, or alkalinity) in the body. This respiratory alkalosis has a braking effect on VE, as the reduced acidity lessens the stimulus to the respiratory control centre to increase VE.

Over a 3- to 5-day period, the excretion of bicarbonate and the retention of hydrogen ions restore the pH of the arterial blood towards normal, and the mechanisms that dampen the VE are lessened.

SUBMAXIMAL HEART RATE AND CARDIAC OUTPUT

Acute exposure to altitude results in an increase in the submaximal heart rate, which reflects an elevated submaximal cardiac output. The increase in cardiac output can compensate for the reduced arterial oxygen content. While the submaximal cardiac output can be increased at altitude, it is not possible to increase the maximal cardiac output. Thus, VO_2max is reduced at altitude. The body considers the increased submaximal cardiac output to be an expensive way of countering the reduced arterial oxygen content. So over the first week spent at altitude, submaximal cardiac output is reduced to sea-level values, or even below sea-level values. During the first week at altitude, there is a progressive increase in the oxygen extraction by the muscles.

Data from the Stenberg et al. (1966) study showed the following:

The cardiac output at 150 watts was 19.2L/min at sea level but 21.6L/min at 4,000m.

Heart rate at 150 watts was 154b/min at sea level and 176b/min at 4,000m.

Points to emphasize:

- With rapid ascent to 4,000m, submaximal cardiac output and heart rate are increased to compensate for the reduced oxygen-carrying capacity of the arterial blood.
- With rapid ascent to 4,000m, the maximum cardiac output is the same at sea level and 4,000m. Thus, it is not possible to compensate for the reduced arterial O_2 content at very high exercise intensities.

HAEMOGLOBIN

There is an increase in haemoglobin concentration at altitude. Initially, this increase is due to a decrease in plasma volume. The 'low oxygen' at altitude is sensed by the kidneys, which release erythropoietin. Over time the erythropoietin increase promotes a greater production of red cells, and an increase in haemoglobin. The increase in haemoglobin is generally advantageous as it allows more O_2 to be carried, but there is also a possible disadvantage: the increase in haemoglobin increases the viscosity of the blood, which can lead to a slowing of the blood flow and a reduction in the cardiac output.

Normal haemoglobin values at sea level are about 15g/100ml for males and about 13.5g/100ml for females. While reinfusion of blood (blood doping) increases the VO_2max at sea level, there is a diminishing benefit at altitude, with no significant increase in VO_2max at 4,300m after blood reinfusion. This is due to the diffusion limitation in the lung. The decrease in PIO_2 at altitude results in a reduced oxygen pressure gradient between the alveoli and venous blood. The diffusion driving force is reduced and the arterial oxygen pressure (PaO_2) is lowered.

ADAPTATIONS AT ALTITUDE

It has been shown that the decrease in PO_2 at altitude results in a decrease in the SaO_2 and a reduced O_2-carrying capacity in the arterial blood. Acclimatization results in an increased VE and SaO_2, although the increases in SaO_2 are not dependent solely on VE but also on other factors. These could include an increased transit time of red cells in the lung resulting from a decreased cardiac output. At altitude (4,300m), resting SaO_2 has been shown to plateau after around 8 days, whereas the exercise SaO_2 increase at altitude has continued to rise between 8 and 22 days of the acclimatization period with no increase in exercise VE. The increased SaO_2 has been shown to enhance endurance performance but not on all occasions.

Comparison of the VE, end tidal CO_2 ($PetCO_2$) and SaO_2, on Day 1 and Day 12 at 4,300m gives an example of changes at altitude (Muza et al. 2004).

At sea level VE was 10.6±0.6L/min

On Day 1 at 4,300m VE was 11.6±0.9L/min

On Day 12 at 4,300m VE was 16.0±1.0L/min

At sea level $PetCO_2$* was 40.0±0.7mmHg

On Day 1 at 4,300m $PetCO_2$ was 28.0±0.6mmHg

On Day 12 at 4,300m $PetCO_2$ was 25.8±0.7mmHg

At sea level SaO_2 was 97±0 per cent

On Day 1 at 4,300m SaO_2 was 84±1 per cent

On Day 12 at 4,300m SaO_2 was 89±1 per cent

The above values were recorded at rest.

* The $PetCO_2$ is the partial pressure of CO_2 (PCO_2) at the end of an expiration and is fairly representative of the PCO_2 in arterial blood.

Points to emphasize:

- the VE is higher at 4,300m than at sea-level. On Day 12 at 4,300m VE was higher than it was on Day 1 at 4,300m;
- the SaO_2 increased over time at altitude;
- the increased SaO_2 results in an enhanced O_2-carrying capacity of the arterial blood.
- the $PetCO_2$ was much lower at altitude as a result of the increased VE.

Altitude acclimatization

Altitude acclimatization results in an increased SaO_2. The increase in SaO_2 has the potential to enhance endurance performance at altitude.

The increase in VE at extreme altitude can be appreciated by noting the resting VE of 40L/min on the summit of Everest compared with a sea-level resting VE of around 8L/min. The estimated $PACO_2$ of 7.5mmHg on the summit of Everest compared with a sea-level value of 40mmHg reflects how the high VE 'blows off' CO_2.

For references, *see* Chapter 6 references.

APPENDIX III

Fluid Balance

The human body can survive for several weeks without food, but death normally ensues in 3–4 days with water deprivation. Water has many important roles in the body, including the transport of foodstuffs in the blood and the excretion of waste products (urine). Water is also vital in the control of core temperature. For example, heat loss through the evaporation of sweat (mainly water) helps cool the body, and water in the body absorbs heat without dramatic changes in temperature.

WATER IN THE BODY

The volume of water in the human body varies depending on the size of the individual, and is related to body composition. Fat contains less water than muscle. Therefore the proportion of fat and muscle will influence how much water is in the body. The 'normal' 70-kg male has 42ltr of water in his body (60 per cent of body weight). Females generally have less water than males as they have slightly more body fat. This water is found in two major compartments: in the cells (the intracellular fluid, or ICF); and outside the cells (the extracellular fluid, or ECF). In the 70-kg man, the ICF makes up 40 per cent of the body weight, at about 28ltr. There are 14ltr of ECF, or 20 per cent of body weight, comprising 10.5ltr of interstitial fluid (fluid that surrounds the cells) and 3.5ltr plasma (the watery part of the blood).

Fluid intake can vary greatly. Water intake for sedentary individuals who live in a temperate climate could be as follows: 1,200ml from drinking, 1,000ml from food, and 350ml as a result of metabolic reactions, giving a total of 2,550ml. Fluid loss may be as follows: 1,500ml from urine, 100ml from faeces, 50ml from sweat and 900ml from insensible fluid loss (for example, through breathing, and seepage of water through the skin), giving a total of 2,550ml. In this example, fluid gain and loss are equal. In some situations this balance will not be evident – for example, an endurance runner may sweat 2ltr per hour for 3 hours and only drink 1ltr of fluid in that time. An individual who lives in and exercises in heat may need to consume 10ltr of fluid per day in order to maintain fluid balance.

REGULATION OF FLUID WITHIN THE BODY

Sensors in the body detect changes in the composition of the extracellular fluid (ECF), and mechanisms are activated to maintain body fluid within certain limits. A number of hormones are involved in this regulation. The hypothalamus in the brain has osmoreceptors that detect changes in the osmolarity (concentration of dissolved particles in a solution) of the extracellular fluid (ECF). An increase in osmolarity results in a higher rate of release of antidiuretic hormone (ADH), which promotes a reduced urine water loss through retention of fluid in the kidneys. In addition, an increase in ADH stimulates thirst and the demand for fluid intake. If the osmolarity of the extracellular fluid (ECF) decreases, the rate of antidiuretic hormone (ADH) release falls, there is a lower perception of thirst and urine output is increased.

The hormone aldosterone regulates sodium absorption. The greater the plasma aldosterone concentration, the greater the retention of sodium in the kidneys will be. An increase in potassium or decrease in sodium in the blood results in an increase in aldosterone. In the body, 'water follows salt'. Thus, if sodium levels are increased, water is retained. Aldosterone also enhances salt sensitivity in the tongue, which may increase body sodium concentration by increasing the intake of salty foods.

Peptides in the heart are released as a result of unusual stretching of the walls of the heart, which comes from an increased blood volume and blood pressure. The release of the peptides reduces the perception of thirst, and lowers the release of antidiuretic hormone (ADH) and aldosterone, thus inhibiting water and salt retention and tending to promote a lower blood volume and blood pressure.

Knowledge of the mechanisms involved in regulating fluid in the body will be of use to the exerciser. For example, the intake of a 'salty' drink at the end of an exercise bout will promote fluid intake and restoration of body fluid levels to normal, as the plasma osmolarity is kept high, resulting in the increased activity of antidiuretic hormone (ADH) and the stimulation of thirst. On the other hand, the intake of plain water will suppress the activity of ADH and lower thirst.

USEFUL WEBSITE ADDRESSES

Travel

Centers for Disease Control and Prevention: www.cdc.gov/
R. Dawood's book *Traveller's Health*: www.travellers-health.info
Dermatology pictures: http://dermatlas.med.jhmi.edu/derm
Health Advice for Travellers, Department of Health UK: www.doh.gov.uk/traveladvice/
Health Canada Travel Medicine Program: www.hc-sc.gc.ca/pphb-dgspsp/tmp-pmv/
International Travel and Health (World Health Organization): www.who.int/ith/
Malaria: www.malaria-reference.co.uk, www.rph.wa.gov.au/labs/haem/malaria/index.html
Malaria test: Optimal IT: www.tcsbiosciences.co.uk
Medical Advisory Services for Travellers Abroad: www.masta.org
National Travel Health Network and Centre: www.nathnac.org
Pictures of parasites and diseases: www.dpd.cdc.gov/dpdx/HTML/image_library.htm
Public Health Laboratories: www.phls.co.uk/facts
Travel Health online: www.tripprep.com
Tropical medicine: www.astmh.org, www.tropmed.tulane.edu/, www.rstmh.org/, www.mahidol.ac.th/mahidol/un/h-tromed.htm
University College London Hospital, Hospital for Tropical Diseases Travel Clinic: www.uclh.org/services/htd/travelclinic.shtml

Altitude

International Society of Travel Medicine: www.istm.org
Medex and Medical Expeditions: www.medex.org.uk
Mountain Medicine and Traumatology Department of Chamonix Hospital: http://Perso.wanadoo.fr/dmtmcham/dmtm_uk.htm
Mountain medicine: www.iemm.org (Institut d'Estudis de Medicina de Muntaya (IEMM)), www.ismmed.org, www.alpinmedizin.org, www.mountainmedicine.ch/
UIAA (Union Internationale des Associations d'Alpinisme/International Mountaineering and Climbing Federation): www.uiaa.ch
UIAA Mountain Medicine Centre: www.thebmc.co.uk/world/mm/mmO.htm
Wilderness Medicine and High Altitude Medicine, Birmingham Medical Research Expeditionary Society: www.bmres.org.uk
Wilderness Medical Society: www.wms.org

Exercise Science and Sports Medicine

American College of Sports Medicine: www.acsm.org (scientific coverage of sport and exercise topics, including fluid replacement guidelines www.acsm.org/publications/newsreleases2005/HyponatraemiaDehydration.htm, and hydration for runners www.acsm.org/publications/newsreleases2005/RunnersHydrate.htm)
Australian Institute of Sport: www.ais.org.au (information on elite sport)
British Association of Sport and Exercise Sciences: www.bases.org.uk/newsite/home.asp (scientific coverage of sport and exercise in the UK)
British Journal of Sports Medicine: http://bjsm.bmjjournals.com/current.shtml (research and review papers on sport, exercise and health)
English Institute of Sport: www.eis2wim.co.uk/gen/ (scientific coverage of elite sport in England)
The Gatorade Sports Science Institute: www.gssiweb.com
Physician and Sports Medicine: www.physsportsmed.com

Scottish Institute of Sport: www.sisport.com
Sport England: www.sportengland.org
Sports science website: www.sportsci.org

Pollution

Air pollution: http://dolphin.upenn.edu/~pennenv/greentime/spring97/air_asin. html; www.lbl.gov/Education/ELSI/pollution-main.html; www.airquality.co.uk/archive/index.php (up-to-date information on regions of UK); http://londonair.org.uk/home.asp (air pollution levels in parts of London)
National Air Quality Information Archives: www.aeat.co.uk/netcen/airqual/
Noise pollution: www.nonoise.org/nipr/wdi98/table3.12.pdf
Pollution: www.worldbank.org (probably the widest source for information on various types of pollution and its intensity and impact in many parts of the world; search the site for different parts of the world and pollution types); www.pollution.com (provides weather forecasts worldwide, and other links); www.scorecard.org (local levels of pollution in all the areas of USA); www.apis.ac.uk/overview/pollutants/overview-03.htm (levels of individual pollutants in the UK, with some maps)
Water pollution: www.umich.edu/-gs265/society/waterpollution.htm

Drug Testing

World Anti Doping Association (WADA): www.wada-ama.org (information on current banned substances, testing procedures and other regulations)

Sports Organizations

International Olympic Committee (IOC): www.olympic.org

REFERENCES

Chapter 1

Gisolfi, C. V. and Wenger, B. C. (1984) Temperature regulation during exercise: old concepts, new ideas, *Exercise and Sports Sciences Reviews*, 12, 339–71
Khogali, M. and Hales, J. R. S. (eds) *Heat Stroke and Temperature Regulation*, Academic Press, Sydney (1983) ISBN 0-12-406180-X
Lloyd, E. L., *Hypothermia and Cold Stress*, Croom Helm (1986) ISBN 0-7099-1665-5
Lloyd, E. L. (1994) ABC of sports medicine: temperature and performance I: cold, *British Medical Journal*, 309, 531–4
McArdle, W. D., Katch, F. I. and Katch, V. L., Exercise and thermal stress. In: *Exercise Physiology, Energy, Nutrition and Human Performance* (Fifth Edition), Lippincott, Williams and Wilkins (2001) ISBN 0 7817-5270-1

Chapter 2

Armstrong, L. E., Costill, D. L. and Fink, W. J. (1985) Influence of diuretic-induced dehydration on competitive running performance, *Medicine Science and Sport and Exercise*, 17, 456–61
Armstrong, L. E., Crago, A. E., Adams, R., et al. (1996) Whole-body cooling of hyper-thermic runners. Comparison of two field therapies, *Journal of American Medicine*, 14, 355–8
Armstrong, L. E., Heat and humidity. In: *Performing in Extreme Environments*, Human Kinetics (2000) ISBN 0-88011-837-7

Armstrong, L. E. (2004) Exertional hyponatraemia, *Journal of Sports Sciences*, 22, 144–7

Cade, J. R., Free, H. J., De Quesada, A. M., et al. (1971) Changes in body fluid composition and volume during vigorous exercise by athletes, *Journal of Sports Medicine and Physical Fitness*, 11, 172–78

Coyle, E. (2004) Fluid and fuel intake during exercise, *Journal of Sports Sciences*, 22, 39–56

Dawood, R., *Travellers' Health*, Oxford University Press (2002) ISBN 0-19-262947-6, *see also* www.travellers-health.info

Edwards, A. M. and Clark, N. A. (2006) Thermoregulatory observations in soccer match play: professional and recreational level applications using an intestinal pill to measure core temperature, *British Journal of Sports Medicine*, 40, 133–8

Fox, E. L., Weiss, H. S., Bartels, R. L., et al. (1966) Thermal responses of man during rest and exercise in a helium oxygen environment, *Archive Environmental Health*, 13, 23–8

Fox, E. L., Bowers, R. W. and Foss, M. L., Exercise in heat and cold. In: *The Physiological Basis of Physical Education and Athletics* (Fourth Edition), Wm C. Brown Publishers, Dubuque, Iowa (1989) ISBN 0-697-05995-2

Gaffin, S. L. and Hubbard, R. W., *Pathophysiology of Heatstroke in Medical Aspects of Harsh Environments, Volume 1, Textbooks of Military Medicine*, Pandolf, K. B. and Burr, R. E. (eds), Offices of The Surgeon General, United States Army, Washington, DC (2001) ISBN 0-16-051184-4

Gaffin, S. L., Gardner, J, W. and Flinn, D. S. (2000) Cooling methods for exertional heat stroke, *Annals of Internal Medicine*, 132, 678–9

Galloway, S. and Maughan, R. (1997) Effects of ambient temperature on the capacity to perform prolonged cycle exercise in man, *Medicine and Science in Sports and Exercise*, 29, 1240–9

Gavin, T. P., Babington, C. A., Harms, M. E., et al. (2001) Clothing fabric does not affect thermoregulation during exercise in moderate heat, *Medicine and Science in Sports and Exercise*, 33, 12, 2124–30

Gonzalez-Alonso, J., Teller, C., Andersen, S. L., et al. (1999) Influence of body temperature on the development of fatigue during prolonged exercise in heat, *Journal of Applied Physiology*, 86, 1032–9

Khogali, M., The Makkah body cooling unit. In: *Heat Stroke and Temperature Regulation* (eds Khogali, M. and Hales, J. R. S.), Academic Press (1983), pp139–48, ISBN 0-12-406180-X

Khogali, M., El-Sayed, H., Amar, M., et al., Management and therapy regimen during cooling and in the recovery room at different heat stroke centres. In: *Heat Stroke and Temperature Regulation* (eds Khogali, M. and Hales, J. R. S.), Academic Press (1983), pp149–56, ISBN 0-12-406180-X

Lloyd, E. L., *Hypothermia and Cold Stress*, Croom Helm (1986), pp121–3, ISBN 0-7099-1665-5

McArdle, W. D., Katch, F. I. and Katch, V. L., Exercise and thermal stress. In: *Exercise Physiology Energy, Nutrition and Human Performance* (Fifth Edition), Lippincott, Williams and Wilkins (2001) ISBN 0-7817-5270-1

MacDougall J. D., Reddan, W. G., Layton, C. R., et al. (1974) Effects of metabolic hyperthermia on performance during heavy prolonged exercise, *Journal of Applied Physiology*, 36, 538–44

Marino, F. E. (2002) Methods, advantages, and limitations of body cooling for exercise performance, *British Journal of Sports Medicine*, 36, 89–94

Marino, F. E., Lambert, M. I. and Noakes, T. D. (2004) Superior performance of African runners in warm humid but not in cool environmental conditions, *Journal of Applied Physiology*, 96, 124–30

Maughan, R. J. (1991) Fluid and electrolyte loss and replacement in exercise, *Journal of Sports Sciences*, 9, 117–42

Maughan, R. J., Merson, S. J., Broad N. P., et al. (2004) Fluid and electrolyte intake and loss in elite soccer players during training, *International Journal of Sport Nutrition and Exercise Metabolism*, 14, 333–46

Maughan, R. and Shirreffs, S. (2004) Exercise in the heat: challenges and opportunities, *Journal of Sports Sciences*, 22, 917–27

Nielsen, B., Hales, J. R. S., Strange, N. J., et al. (1993) Human circulatory and thermoregulatory adaptions with heat acclimation and exercise in a hot, dry environment, *Journal of Physiology*, 460, 467–85

Noakes, T., *The Lore of Running*, Human Kinetics, Champaign, Illinois (2001) ISBN 0-87322-959-2

Noakes, T., Sharwood, K., Collins, M., et al. (2004) The dipsomania of great distance: water intoxication in an Ironman triathlete, *British Journal of Sports Medicine*, 38e, 16

Noakes, T. and Speedy, D. (2006) Case proven: exercise associated hyponatraemia is due to overdrinking. So why did it take 20 years before the original evidence was accepted? *British Journal of Sports Medicine*, 40, 567–72

Sawka, M. N. and Wenger, C. B., Physiological responses to acute exercise-heat stress. In: *Human Performance Physiology and Environmental Medicine at Terrestrial Extremes* (eds Pandolf, K., Sawka, M. N. and Gonzales, R. R.), Benchmark Press, Indianapolis (1988) ISBN 1-884125-02-6

Sharwood, K. A., Collins, M., Goedecke, J. H. et al. (2004) Weight changes, medical complications, and performance during an Ironman triathlon, *British Journal of Sports Medicine*, 38, 718–24

Shirreffs, S. M., Aragon-Vargas, L. F., Chamorro, M., et al. (2005) The sweating response of elite professional soccer players to training in the heat, *International Journal of Sports Medicine*, 26, 90–5

Sigurbjorn, A., Arngrimsson, D. S., Petitt, M. G., et al. (2004) Cooling vests worn during active warm-up improves 5-km run performance in the heat, *Journal of Applied Physiology*, 96, 1867–74

Tappan, D. V., Jacey, M. J., Heyder, E. and Gray, P. H. (1984) Blood volume responses in partially dehydrated subjects working in the cold, *Aviation, Space and Environmental Medicine*, 55, 296–301

Taylor, S. and Diffey, B. (2002) Simple doseage guide for suncreams will help users, *British Medical Journal*, 324, 1526

Walpole, B., Noakes, T. D. and Collins, M. (2006) Growth hormone 1 (GH1) gene and performance and post-race rectal temperature during the South African Ironman triathlon, *British Journal of Sports Medicine*, 40, 145–50

Chapter 3

Almond, C. S. D., Shin, A. Y. and Fortescue, E. B. (2005) Hyponatraemia among runners in the Boston Marathon, *New England Journal of Medicine*, 352, 1550–6

American College of Sports Medicine (1996) Position Stand on Exercise and Fluid Replacement, *Medicine and Science in Sport and Exercise*, 28, i–vii

American College of Sports Medicine (2005) Guidance to athletes on preventing hyponatraemia and dehydration during races. Newly published roundtable statement on hydration and physical activity reinforce importance of managing both conditions. *See also* www.acsm.org/publications/newsreleases2005/Hyponatraemia Dehydration.htm

Armstrong, L. (2004) Exertional hyponatraemia, *Journal of Sports Sciences*, 22, 144–6

Armstrong, L. E. (2006) Commentary 4, *British Journal of Sports Medicine*, 40, 106

Burke, L. (1997) Fluid balance during team sports, *Journal of Sports Sciences*, 15, 287–95

Burke, L. M., Kiens, B., Ivy, J. L., et al. (2004) Carbohydrates and fat for training and recovery, *Journal of Sports Sciences*, 22, 15–30

Costill, D. L. and Hargreaves, M. (1992) Carbohydrate nutrition and fatigue, *Sports Medicine*, 13, 86–92

Coyle, E. (2004) Fluid and fuel intake during exercise, *Journal of Sports Sciences*, 22, 39–56

Dugas, J. P., Noakes, T. D. and Laursen, P. B. (2005) Hyponatraemic encephalopathy despite a modest rate of fluid intake during a 109km cycle race, *British Journal of Sports Medicine*, 39, e38

Friggell, R. T., Lang, G. H., Lowance, D. E. et al. (1986) Hyponatraemia and ultra-marathon running. *Journal of the American Medical Association*, 255, 772–74

Gavin, T. P., Babington, P. J., Harms, C. A., et al. (2001) Clothing fabric does not affect thermoregulation during exercise in moderate heat, *Medicine and Science in Sports and Exercise*, 33, 2124–230

Gisolfi, C. V. and Duchman, S. M. (1992) Guidelines for optimal replacement beverages for different athletic events, *Medicine and Science in Sports and Exercise*, 24, 679–87

Hew-Butler, T. D., Sharwood, K., Collins, M. et al. (2006) Sodium supplementation is not required to maintain serum sodium concentrations during an Ironman triathlon. *British Journal of Sports Medicine*, 40, 255–59

Hsieh, M. (2004) Recommendations for treatment of hyponatraemia at endurance events, *Sports Medicine*, 34, 231–8

Maughan, R., Fenn, C. and Leiper, J. (1989) Effects of fluid, electrolyte and substrate ingestion on endurance capacity, *European Journal of Applied Physiology*, 58, 481–6

Maughan, R. J. (1991) Fluid and electrolyte loss and replacement in exercise, *Journal of Sports Sciences*, 9, 117–42

Maughan, R. and Leiper, J. (1995) Sodium intake and post-exercise rehydration in man, *European Journal of Applied Physiology*, 71, 311–19

Maughan, R., Shirreffs, S. M., Merson, S. J., et al. (2005) Fluid and electrolyte balance in elite male football (soccer) players training in a cool environment, *Journal of Sports Sciences*, 23, 73–9

Maughan, R., Sports drinks before, during and after running. In: *Proceedings of Marathon Medicine 2000 London*, The Royal Society of Medicine, 2001, pp147–58

Montain, S. and Coyle, E. (1992) Benefits of fluid replacement with carbohydrate during exercise, *Medicine and Science in Sports and Exercise*, 24, S324–30

Montain, S. and Coyle, E. (1992) Influence of graded dehydration on hyperthermia and cardiovascular drift during exercise, *Journal of Applied Physiology*, 73, 1340–50

Montain, S. J., Cheuvront, S. N. and Sawka, M. N. (2006) Exercise-associated hyponatraemia: quantitative analysis to understand the aetiology, *British Journal of Sports Medicine*, 40, 98–106

Murray, R. (2006) Commentary 3, *British Journal of Sports Medicine*, 40, 106

Noakes, T. D. (1981) Comrades make medical history – again. *South African Runner*, 4, 8–10

Noakes, T., *The Lore of Running*, Human Kinetics, Champaign, Illinois (2001) ISBN 0-87322-959-2

Noakes, T., Sharwood, K., Collins, M., et al. (2004) The dipsomania of great distance: water intoxication in an Ironman triathlete, *British Journal of Sports Medicine*, 38e, 16

Noakes, T. (2006) Commentary 1, *British Journal of Sports Medicine*, 40, 105

Noakes, T. and Speedy, D. (2006) Case proven: exercise associated hyponatraemia is due to overdrinking. So why did it take 20 years before the original evidence was accepted? *British Journal of Sports Medicine*, 41, 567–72

Pascoe, D. D., Shanley, L. A. and Smith, E. W. (1994) Clothing and exercise: 1. Biophysics of heat transfer between the individual, clothing and environment, *Sports Medicine*, 18, 38–54

Pascoe, D. D., Shanley, L. A. and Smith, E. W. (1994) Clothing and exercise: 2. Influence of clothing during exercise/work in environmental extremes, *Sports Medicine*, 18, 94–108

Rehrer, N. (2001) Fluid and electrolyte balance in ultra-endurance sport, *Sports Medicine*, 31, 701–15

Shirreffs, S., Armstrong, L., Cheuvront, S. (2004) Fluid and electrolyte needs for preparation and recovery from training and competition, *Journal of Sports Sciences*, 22, 57–64

Twerenbold, R., Knechtle, B., Kakebeeke, T. H., et al. (2003) Effects of different sodium concentrations in replacement fluids during prolonged exercise in women, *British Journal of Sports Medicine*, 37, 300–03

Weschler, L. B. (2006) Commentary 2, *British Journal of Sports Medicine*, 40, 105–06

Chapter 4

Adams, W. C., Fox, R. H., Fry, A. J. and MacDonald, I. C. (1975) Thermoregulation during marathon running in cool, moderate and hot environments, *Journal of Applied Physiology*, 38, 1030–7

Fox, E. L., Bowers, R. W. and Foss, M. L., Exercise in heat and cold. In: *The Physiological Basis of Physical Education and Athletics* (Fourth Edition), Wm C. Brown Publishers, Dubuque, Iowa (1989) ISBN 0-697-05995-2

Houmard, J. A., Costill, D. L., Davis, J. A. et al. (1990) The influence of exercise intensity on heat acclimation in trained subjects, *Medicine and Science in Sports and Exercise*, 22, 615–20

Lind, A. and Bass, D. (1963) Optimal exposure time for development of acclimatization to heat, *Federation Proceedings*, 22, 704–8

McArdle, W. D., Katch, F. I. and Katch, V. L., Exercise and thermal stress. In: *Exercise Physiology, Energy, Nutrition and Human Performance* (Fifth Edition), Lippincott, Williams and Wilkins (2001) ISBN 0-7817-5270-1

Montain, S. J., Maughan, R. J. and Sawka, M. N. (1996) Heat acclimatization strategies for the 1996 Summer Olympics, *Athletic Therapy Today*, 1, 42–6

Nadel, E. R., Pandolf, K. B., Roberst, M. F., et al. (1974) Mechanisms of thermal acclimation to exercise and heat, *Journal of Applied Physiology*, 37, 515–20

Nielsen, B., Hales, S., Strange, N., et al. (1993) Human circulatory and thermoregulatory adaptations with heat acclimation and exercise in a hot, dry environment, *Journal of Physiology* (Lond) 460, 467–85

Noakes, T., *The Lore of Running*, Human Kinetics, Champaign, Illinois (2001) ISBN 0-87322-959-2

Pichan, G., Sridharan, K. and Swamy, Y. V. (1985) Physiological acclimatization to heat after a spell of cold conditioning in tropical subjects, *Aviation Space and Environmental Medicine*, 56, 436–40

Sawka, M. N., Toner, M. M., Francesconi, R. P., et al. (1983) Hypohydration and exercise: effects of heat acclimation, gender and environment, *Journal of Applied Physiology*, 55, 1147–53

Sawka, M. N. and Wenger, C. B., Physiological responses to acute exercise-heat stress. In: *Human Performance Physiology and Environmental Medicine at Terrestrial Extremes* (eds Pandolf, K., Sawka, M. N. and Gonzales, R. R.), Benchmark Press, Indianapolis (1988) ISBN 1-884125-02-6

Sawka, M. N. and Pandolf, K. B., Physical exercise in hot climates: physiology, performance and biomedical issues. In: *Medical Aspects of Harsh Environments, Volume 1,*

Textbooks of Military Medicine (eds Pandolf, K. B. and Burr, R. E.), Offices of The Surgeon General, United States Army, Washington, DC (2001) ISBN 0-16-051184-4

Wenger, B. C. Adaption to hot environments. In: *Human Performance Physiology and Environmental Medicine at Terrestrial Extremes* (eds Pandolf, K., Sawka M. N. and Gonzales, R. R.), Benchmark Press, Indianapolis (1986) ISBN 1-884125-02-6

Williams, C. G., Wyndham, C. H. and Morrison, J. F. (1967) Rate of loss of acclimatization in summer and winter, *Journal of Applied Physiology*, 22, 21–6

Young, Human adaptation to cold, p402. In: *Human Performance Physiology and Environmental Medicine at Terrestrial Extremes* (eds Pandolf, K., Sawka M. N. and Gonzales, R. R.), Benchmark Press, Indianapolis (1986) ISBN 1-884125-02-6

Chapter 5

Davies, M., Ekblom, B., Bergh, U., et al. (1975) The effects of hypothermia on submaximal and maximal work performance, *Acta Physiol Scand*, 95, 201–2

Dickinson, A. L., Wilderness clothing. In: *Wilderness Medicine*, Mosby Publishing, St Louis, Missouri (1995) ISBN 0-8016-7044-6

Doubt, T. J. (1991) Physiology of exercise in the cold, *Sports Medicine*, 11, 367–81

Fox, E. L., Bowers, R. W. and Foss, M. L., Exercise in heat and cold. In: *The Physiological Basis of Physical Education and Athletics* (Fourth Edition), Wm C. Brown Publishers, Dubuque, Iowa (1989) ISBN 0-697-05995-2

Galloway, S. and Maughan, R. (1997) Effects of ambient temperature on the capacity to perform prolonged cycle exercise in man, *Medicine Science and Sport and Exercise*, 29, 1240–9

Gilbert, M., Busund, R., Skagseth, A., et al. (2000) Resuscitation from accidental hypothermia of 13.7°C with circulatory arrest, *The Lancet*, 355, 375–6

Harnett, R. M., Pruitt, J. R. and Sias, F.R. (1983) A review of the literature concerning resuscitation from accidental hypothermia: Part 1 – The problem and general approaches, *Aviation. Space and Environmental Medicine*, 54, 425–34

Lahti, A. (1982) Cutaneous reactions to cold, *Nordic Council Arctic Medicine Research*, Rep No 30, 32–5

Laufmann, H. (1951) Profound accidental hypothermia, *Journal of American Medical Assocciation*, 147, 1201–12

Lloyd, Evan L., *Hypothermia and Cold Stress*, Croom Helm, London (1986) ISBN 0-7099-1665-5

Lloyd, E. (1994) ABC Sports Medicine, *British Medical Journal*, 309, 531–4

Lloyd, Evan L. (1996) Accidental hypothermia, *Resuscitation* 32, 111–24

Lowdon, B. J. (1984) Hazards of cold-water swims in Triathlete events, *Official Gazette of Australian Sports Medicine Federation*, 2, 17–19

McArdle, W. D., Katch, F. I. and Katch, V. L., Exercise and thermal stress. In: *Exercise Physiology, Energy, Nutrition and Human Performance* (Fifth Edition), Lippincott, Williams and Wilkins (2001) ISBN 0-7817-5270-1

Nadel, E. R., Holmer, I. Bergh, U., et al. (1974) Energy exchanges of swimming man, *Journal of Applied Physiology*, 36, 465–71

Niazi, S. A. and Lewis, F. J. (1957) Profound hypothermia in monkey with recovery after long periods of cardiac standstill, *Journal of Applied Physiology*, 10, 137–8

Niazi, S. A. and Lewis, F. J. (1958) Profound hypothermia in man; report of a case, *Annals of Surgery*, 147, 254–66

Noakes, T. D. (2000) Exercise and the cold, *Ergonomics*, 43, 1461–79

Noakes, T., *The Lore of Running*, Human Kinetics, Champaign, Illinois (2001) ISBN 0-87322-959-2

Nozaki, R. N., Ishabashi, K., Adachi, N. (1986) Accidental profound hypothermia, *New England Journal of Medicine*, 315, 1680

Pascoe, D. D., Shanley, L. A. and Smith, E. W. (1994) Clothing and exercise: 1. Biophysics of heat transfer between the individual, clothing and environment, *Sports Medicine*, 18, 38–54

Pascoe, D. D., Shanley, L. A. and Smith, E. W. (1994) Clothing and exercise: 2. Influence of clothing during exercise/work in environmental extremes, *Sports Medicine*, 18, 94–108

Patton, J. F. and Vogel, J. A. (1984) Effects of acute cold exposure on submaximal endurance performance, *Medicine and Science in Sports and Exercise*, 16, 494–7

Prendergast, D. R. (1988) The effect of body cooling on oxygen transport during exercise, *Medicine and Science in Sports and Exercise*, 20, S171–S176

Pugh, G. (1967) Cold stress and muscular exercise, with special reference to accidental hypothermia, *British Medical Journal*, 2, 333–7

Pugh, G. G. C. E. (1967) Cold stress and muscular exercise, with special reference to accidental hypothermia, *British Medical Journal*, 2, 333–7

Pugh, L. G. C. E. and Edholm, O. G. (1955) The physiology of Channel swimmers, *Lancet*, 2, 761–7

Smith, A. U. (1959) Viability of supercooled and frozen mammals, *Annals of New York Academy of Sciences*, 80, 291–300

Tipton, M., Eglin, C., Gennser, M. and Golden, F. (1999) Immersion deaths and deterioration in swimming performance in cold water, *The Lancet*, 354, 626–9

Toner, M. M. and McArdle, W. D., Physiological adjustments of man to the cold. In: *Human Performance Physiology and Environmental Medicine at Terrestrial Extremes* (eds Pandolf, K. B., Sawka, M. N. and Gonzalez, R. R.), Cooper Publishing Group, USA (1986) ISBN 1-884125-02-6

Webb, P. (1951) Air temperature in respiratory tracts of resting subjects in the cold, *Journal of Applied Physiology*, 4, 378–82

Wood, V. (1977) Case of hypothermia, *Canadian Medical Association*, 117, 16–17

Chapter 6

Adams, W. C., Bernauer, E. M., Dill, D. B., et al. (1975) Effects of equivalent sea-level and altitude training on VO_2 max and running performance, *Journal of Applied Physiology*, 39, 262–6

Balke, B., Nagle, J. and Daniels, J. (1965) Altitude and maximum performance in work and sports activity, *Journal of American Medical Association*, 194, 646–9

Basnyat, B. and Murdoch, D. R. (2003) High-altitude illness, *The Lancet*, 361, 1967–1974

Buskirk, E. R., Kollias, J., Akers, R. F., Prokop, E. K. and Reatagui, E. P. (1967) Maximal performance at altitude and on return from altitude in conditioned runners, *Journal of Applied Physiology*, 23, 774–6

Chapman, R. F., Stray-Gundersen, J. and Levine, B. D. (1998) Individual variation in response to altitude training, *Journal of Applied Physiology*, 85, 1448–56

Daniels, J. and Oldridge, N. (1970) The effects of alternate exposure to altitude and sea level on world-class middle-distance runners, *Medicine and Science in Sports and Exercise*, 2, 107–12

Faulkner J. A., Daniels, J. T. and Balke, B. (1967) Effects of training at moderate altitude on physical performance capacity, *Journal of Applied Physiology*, 23, 85–9

Faulkner, J. A., Kollias, J., Favour, C., et al. (1968) Maximum aerobic capacity and running performance, *Journal of Applied Physiology*, 24, 685–91

Fulco, C. S., Rock, P. B. and Cymerman, A. (2000) Improving athletic performance: is altitude residence or altitude training helpful? *Aviation Space Environmental Medicine*, 71, 162–71

Fulco, C. S. and Cymerman, A., Human performance and acute hypoxia. In: *Human Performance Physiology and Environmental Medicine at Terrestrial Extremes* (eds

Pandolf, K. B., Sawka, M. N. and Gonzalez, R. R.), Brown and Benchmark, USA (1986) ISBN 1-884125-02-6

Gore, C. J., Little, S. C. and Hahn, A. G. (1997) Reduced performance of male and female athletes at 580m altitude, *European Journal of Applied Physiology*, 75, 136–43

Gore, C. J., Craig, N., Hahn, A. G., et al. (1998) Altitude training at 2690m does not increase total haemoglobin mass or sea-level VO_2max in world champion track cyclists, *Journal of Science and Medicine in Sport*, 1, 156–70

Hahn, A. G. and Gore C. J. (2001) The effect of altitude on cycling performance: a challenge to traditional concepts, *Sports Medicine*, 31, 533–57

Levine, B. D. and Stray-Gundersen, J. (1997) Living high and training low; effect of moderate-altitude acclimatization with low-altitude training on performance, *Journal of Applied Physiology*, 83, 102–112

Levine, B. D. and Stray-Gunderson, J., The effects of altitude training are mediated primarily by acclimatization, rather than by hypoxic exercise: Chapter 7 pp75–88. In: *Hypoxia: From genes to the bedside. Advances in Experimental Medicine and Biology* (eds Roach, R. C., Wagner, P. D. and Hackett, P. H.), Kluver Academic/ Plenum Publishers, New York (2001) ISBN-0-306-46696-1

McArdle, W. D., Katch, F. I. and Katch, V. L., Exercise at medium and high altitude. In: *Exercise, Physiology Energy, Nutrition and Human Performance* (Fifth Edition), Lippincott, Williams and Wilkins (2001) ISBN 0-7817-5270-1

McFarland, R, A. and Dill, D. B. (1938) A comparative study of the effects of reduced oxygen pressure on man during acclimatization, *Journal of Aviation Medicine, 9*, 18–34

Mizuno, M., Juel, C., Bro-Rasmussen, T., et al. (1990) Limb skeletal muscle adaptation in athletes after training at altitude, *Journal of Applied Physiology*, 68, 496–502

Muza, S. R., Young, A. J., Sawka, M. N., et al. (2004) Ventilation after supplemental oxygen administration at high altitude, *Wilderness & Environmental Medicine*, 15, 18–24

Noakes, T., *The Lore of Running*, Human Kinetics, Champaign, Illinois (2001) ISBN 0-87322-959-2

Numella, A. T. and Rusko, H. (2000) Acclimatization to altitude and normoxic training improve 400m running performance at sea level, *Journal of Sports Sciences*, 18, 411–19

Pugh, L. G. C. E. (1967) Athletes at altitude, *Journal of Physiology*, 192, 619–46

Ri-Li Ge, S., Witkowski, Y., Zhang, C., et al. (2002) Determinants of erythropoieten release in response to short-term hypobaric hypoxia, *Journal of Applied Physiology*, 92, 2361–7

Roberts, A. D., Clark, S. A., Townsend, et al. (2003) Changes in performance, maximal oxygen uptake ands maximum accumulated oxygen deficit after 5, 10 and 15 days of live high:train low altitude exposure, *European Journal of Applied Physiology*, 88, 390–5

Roels, B., Hellard, P., Schmitt, L., et al. (2006) Is it more effective for highly trained swimmers to live and train at 1200m than at 1850m in terms of performance and haematological benefits? *British Journal of Sports Medicine*, 40, e4

Rusko, H., Tikkanen, H. and Peltonen, J. (2004) Altitude and endurance training, *Journal of Sports Sciences*, 22, 928–45

Saunders, P. U., Telford, R. D., Pyne, D. B., et al. (2004) Improved running economy in elite runners after 20 days of simulated moderate-altitude exposure, *Journal of Applied Physiology*, 96, 931–7

Smith, M. H. and Sharkey, B. J. (1984) Altitude training: who benefits? *Physician and Sports Medicine*, 12, 48–62

Stenberg, J., Ekblom, B. and Messin, R. (1966) Hemodynamic response to work at simulated altitude, 4000m, *Journal of Applied Physiology*, 21, 1589–94
Ward, M. P., Milledge, J. S. and West, J. B., *High Altitude Medicine and Physiology* (Third Edition), Arnold, London (2000) ISBN 0-340-75980-1
Wehrlin, J. P. and Marti, B. (2006) Live high–train low associated with increased haemoglobin mass as preparation for the 2003 World Championships in two native European world class runners, *British Journal of Sports Medicine*, 40, e3
Young, A. J. and Young, P. M., Human acclimatization to high terrestrial altitude. In: *Human Performance Physiology and Environmental Medicine at Terrestrial Extremes* (eds Pandolf, K. B., Sawka, M. N. and Gonzalez, R. R.), Brown and Benchmark, USA (1986) ISBN 1-884125-02-6

Chapter 7

Armstrong, L. E., Air pollution: exercise in the city. In: *Performing in Extreme Environments*, Human Kinetics (2000) ISBN 0-88011-837-7
Carlisle, A. J. and Sharp, N. C. C. (2001) Exercise and outdoor ambient pollution, *British Journal of Sports Medicine*, 35, 214–22
Colls, J., *Air Pollution, An Introduction*, E & FN Spon, 1997, ISBN 0-41-920650-7
Dawood, R., *Travellers' Health*, Oxford University Press (2002) ISBN 0-19-262947-6, *see also* www.travellers-health.info
Florida-James, G., Donaldson, K and Stone, V. (2004) Athens 2004: the pollution climate and athletic performance, *Journal of Sports Sciences*, 22, 967–80
Folinsbee, L. and Raven, P. R., Air pollution – acute and chronic effects. In: *Proceedings of Marathon Medicine 2000*, London, The Royal Society of Medicine (2001)
Foxcroft, W. J. and Adams, W. C. (1986) Effects of ozone on four consecutive days on work performance and VO_2max, *Journal of Applied Physiology*, 61, 960–6
Gong, H., Health effects of air pollution. A review of clinical studies, *Clinics in Chest Medicine*, 13, 201–14
Gong, H. and Krishnareddy, S. (1995) How pollution and airborne allergens affect exercise, *Physician and Sports Medicine*, 23, 35–42
Hage, P. (1982) Air pollution: adverse effect on performance, *Physician and Sports Medicine*, 10, 127–32
Horner, J. (1998) Carbon monoxide: the invisible killer, *Journal of Royal Society of Health*, 118, 141–5
Horvath, S. M., Raven, P. B., Dahms, T. E., et al. (1975) Maximal aerobic capacity at different levels of carboxyhemoglobin, *Journal of Applied Physiology*, 38, 300–03
Horvath, S. M. (1981) Impact of air quality in exercise performance, *Exercise and Sports Science Reviews*, 9, 265–96
Leavitt, F. J. and Varanase, M. (2005) Making dirty water drinkable, *British Medical Journal*, 331, 781
Lee, K., Yanagisawa, Y., Spengler, J. D., et al. (1994) Carbon monoxide and nitrogen dioxide in indoor ice skating rinks, *Journal of Sports Sciences*, 12, 279–83
Linn, W. S., Venet, T. G., Shamoo, D. A., et al. (1983) Respiratory effects of sulphur dioxide in heavily exercising asthmatics, *American Review of Respiratory Disease*, 127, 278–83
MacSween, R. M., Williams, H. C. (1996) Seabather's eruption – a case of Caribbean itch, *Br Med J*. 312, 957–8
Neumann, K. (1999) Staying healthy in Mexico, *Wilderness Med Letter*, 16, No. 3, p178
Nicholson, J. P. and Case, D. B. (1983) Carboxyhemoglobin levels in New York City runners, *Physician and Sports Medicine*, 1, 135–8
Pandolf, K. B., Air quality and human performance. In: *Human Performance Physiology and Environmental Medicine at Terrestrial Extremes* (eds Pandolf, K. B., Sawka, M.

N. and Gonzalez, R. R.), Brown and Benchmark, USA (1986) ISBN 1-884125-02-6

Schelegle, E. S. and Adams, W. C. (1986) Reduced exercise time in competitive simulations consequent to low-level ozone exposure, *Medicine and Science in Sports and Exercise*, 18, 408–14

Speizer, F. and Frankj, A. (1966) A comparison of changes in pulmonary flow resistance in healthy volunteers acutely exposed to SO_2 by mouth and nose, *British Journal of Independent Medicine*, 23, 73–9

Sterling, T. D., Dimich, H. and Kobayashi, D. (1982) Indoor by-product levels of tobacco smoke: a critical review of the literature, *Journal of Air Pollution Control Association*, 29, 238–41

Chapter 8

Armstrong, L. E., Biorhythmic disturbances. In: *Performing in Extreme Environments*, Human Kinetics (2000) ISBN 0-88011-837-7

Atkinson, G. and Reilly, T. (1996) Circadian variation in sports performance, *Sports Medicine*, 21, 292-311

Auerbach, Paul S. (ed) *Wilderness Medicine: Management of Wilderness and Environmental Emergencies* (Third Edition), Mosby-Year Book, St Louis (1995) ISBN 0-8016-7044-6

Buscemi, N., Vandermeer, B., Hooton, P. R., et al., *Melatonin for treatment of sleep disorders*, Rockville, M.D.; Agency for Healthcare Research and Quality, 2004. (Evidence report/technology assessment No 108, AHRQ publication No 05-E002-1.)

Canadian Medical Association Journal (1996) Travel statement on jet lag, *Canadian Medical Association Journal*, 155, 61–3

Dawood, R., *Traveller's Health*, Oxford University Press (2002) ISBN 0-19-262947-6, *see also* www.travellers-health.info

Herxheimer, A. and Waterhouse, J. (2003) The prevention and treatment of jet lag: it's been ignored, but much can be done, *British Medical Journal*, 326, 296–7

Horton, P. (1997) Redback spider is now established in Japan: bites can be recognised by a unique sign, *British Medical Journal*, 314, 1484

Kerkhoffs, G. M. M. J., op den Akker, J. W. and Hammacher, E. R. (2003) Surfer wipe-out by predator fish, *British Journal of Sports Medicine*, 37, 537–9

NewGenn Research Limited, The Vision Centre, Eastern Way, Bury St Edmunds, Suffolk, England IP32 7AB (www.newgenn.com)

Pollard, A. J. and Murdoch, D. R., *The High-Altitude Medicine Handbook*, Radcliffe Medical Press Ltd, Oxford (2003) ISBN 1-85775-849-8

Reilly, T. and Waterhouse, J., *The Stress of Travel in Sport, Exercise and Environmental Physiology*, Elsevier, London (2005) ISBN 0-443-07358-0

Walker, E., Williams, G. and Raeside, F., *ABC of Healthy Travel* (Fourth Edition), BMJ Publishing Group (1993) ISBN 0-7279-0762-X

Warrell, D. A. (2005) Treatment of bites by adders and exotic venomous snakes, *British Medical Journal*, 331, 1244–47

Waterhouse, J., Reilly, T. and Atkinson, G. (1997) Jet lag, *Lancet*, 350, 1611–15

Waterhouse, J., Edwards, B. and Nevill, A. (2002) Identifying some determinants of 'jet lag' and its symptoms: a study of athletes and other travellers, *British Journal of Sports Medicine*, 36, 54–60

Waterhouse, J., Reilly, T. and Edwards, B. (2004) The stress of travel, *Journal of Sports Sciences*, 22, 946–66

Winget, C. M., DeRoshia, C. W., Markley, B. A., et al. (1984) A review of human physiological and performance changes associated with desynchronosis of biological rhythms, *Aviation, Space and Environmental Medicine*, 55, 1085–96

Winget, C. M., De Roshia, C. W. and Holley D. C. (1985) Circadian rhythms and athletic performance, *Medicine and Science in Sports and Exercise*, 17, 498–516

Chapter 9

Ayres, J. (1994) Asthma and the atmosphere, *British Medical Journal*, 309, 619–20

Celenza, A., Fothergill, J., Kupek, E. and Shaw, R. J. (1996) Thunderstorm associated asthma: a detailed analysis of environmental factors, *British Medical Journal*, 312, 604–7

Dawood, R., *Traveller's Health*, Oxford University Press (2002) ISBN 0-19-262947-6, Appendix 10, p673, *see also* www.travellers-health.info

Dickinson, J. W., Whyte, G. P., McConnell, A. K. and Harries, M. G. (2006) Screening elite winter athletes for exercise induced asthma: a comparison of three challenge methods, *British Journal of Sports Medicine*, 40, 179–83

Herxheimer, A. and Waterhouse, J. (2003) The prevention and treatment of jet lag: it's been ignored, but much can be done, *British Medical Journal*, 326, 296–7

Kavanagh, T., *Heart Attack? Counter Attack!* Van Nostrand Reinhold Ltd, Toronto (1976) ISBN 0-442-29927-3

Larsson, K., Ohlsen, P., Larsson, L., et al. (1993) High prevalence of asthma in cross-country skiers, *British Medical Journal*, 307, 1326–9

Lloyd, E. L., *Hypothermia and Cold Stress*, Croom Helm, London (1986) ISBN 0-7099-1665-5

Moore, K., Thompson, C., Hayes, R. (2001) Diabetes and extreme altitude mountaineering. *British Journal of Sports Medicine*, 35, 85

Nadel, J. A. and Widdicombe, J. G. (1962) Effects of changes in blood gas tensions and carotid sinus pressure on tracheal volume and total lung resistance to airflow, *Journal of Physiology (Lond)*, 163, 13–33

Newhouse, M. T., Becklake, M. R., Macklem, P. T., et al. (1964) Effect of end-tidal CO_2 tension on flow resistance, *Journal of Applied Physiology*, 19, 745–9

Pollard, R. C. and Pollard, A. J. (1994) Asthma in cross-country skiers, *British Medical Journal*, 308, 274

Sterling, G. M., The mechanism of bronchoconstriction due to hypoxia in man, *Clinical Science*, 35, 105–14

Thames Regions Accident and Emergency Trainees Association, et al. (1996) A major outbreak of asthma associated with a thunderstorm: experience of accident and emergency departments and patients' characteristics, *British Medical Journal*, 312, 601–4

Tikkanen, H. O. and Helenius, I. (1994) Asthma in runners, *British Medical Journal*, 309, 1087

Voy, R. O. (1986) The US Olympic Committee experience with exercise-induced bronchospasm, *Medicine Science and Sports in Exercise*, 18, 328–30

Chapter 10

Alonso, P. L., Sacarial, J., Aponte, J. J., et al. (2005) Duration of protection with RTS,S/AS02A malaria vaccine in prevention of *Plasmodium falciparum* disease in Mozambican children: single-blind extended follow-up of a randomized controlled trial, *Lancet*, 366, 2012–18

Chanteau, S., Rahalison, L., Ralafiarisoa, L., et al. (2003) Development and testing of a rapid diagnostic test for bubonic and pneumonic plague, *Lancet*, 361, 211–16

Dawood, R., *Travellers' Health*, Oxford University Press (2002) ISBN 0-19-262947-6, *see also* www.travellers-health.info

Dobson, M. (1996) Vaccines, malaria, and a host of resistance, *British Medical Journal*, 313, 67–8

Gillespie, S. (2001) Toxicara: dogwalking and playing fields, *British Medical Journal*, 35, 6–7
Kordi, R. and Wallace, W. A. (2004) Blood-borne infections in sport: risks of transmission, methods of prevention, and recommendations for hepatitis B vaccination, *British Medical Journal*, 38, 678–84
Moody, A. H., Chiodini, P. L. (2002) Non-microscopic method for malaria diagnosis using OptiMAL IT, a second-generation dipstick for malaria pLDH antigen detection, *British Journal of Biomedical Science*, 59, 228–31
Nadelman, R. B., Nowakowski, J., Fish, D., et al. (2001) Prophylaxis with single dose doxycycline for the prevention of Lyme disease after an *Ixodes scapularis* tick bite, *New England Journal of Medicine*, 345, 79–84
Pearn, J. H. (1997) Chronic fatigue syndrome: chronic ciguatera poisoning as a differential diagnosis, *Medical Journal of Australia*, 166, 309
Pounder, D. (2005) Avoiding rabies, *British Medical Journal*, 331, 469–70
Solomon, T., Marston, D., Mallewa, M., et al. (2005) Paralytic rabies after a two-week holiday in India, *British Medical Journal*, 331, 501–3
Young, D. A. B. (1995) Florence Nightingale's fever, *British Medical Journal*, 311, 1711–14

Appendix I

American College of Sports Medicine, *Guidelines for graded exercise testing and Exercise Prescription*, Lea & Febiger (1986) ISBN 0-8121-1022-6
McArdle, W. D., Katch, F. I. and Katch, V. L., *Exercise Physiology, Energy, Nutrition and Human Performance* (Fifth Edition), Lippincott, Williams and Wilkins (2001) ISBN 0-7817-5270-1
McFarland, R, A. and Dill, D. B. (1938) A comparative study of the effects of reduced oxygen pressure on man during acclimatization, *Journal of Aviation Medicine*, 9, 18–34
Muza, S. R., Young, A. J., Sawka, M. N., et al. (2004) Ventilation after supplemental oxygen administration at high altitude, *Wilderness & Environmental Medicine*, 15, 18–24
Stenberg, J., Ekblom, B., Messin, R. (1966) Hemodynamic response to work at simulated altitude, 4000m, *Journal of Applied Physiology*, 21, 1589–94
Vander, A. J., Sherman, J. H. and Luciano, D. S., *Human Physiology: The Mechanisms of Body Function* (Sixth Edition), McGraw-Hill, Inc., New York (1994) ISBN 0-07-Y66619-9 (later editions also available)

Appendix II

See References for Chapter 6

FURTHER READING

Auerbach, Paul S. (ed), *Wilderness Medicine: Management of Wilderness and Environmental Emergencies* (Third Edition), Mosby-Year Book, St Louis (1995) ISBN 0-8016-7044-6
Dawood, R., *Travellers' Health*, Oxford University Press (2002) ISBN 0-19-262947-6, *see also* www.travellers-health.info
Geddes, A. M. (1994) Zoonoses in the United Kingdom, *Proceedings Royal College of Physicians of Edinburgh*, 24, 534–9
Kumar, A. and Kumar, C. J. (2005) . . . as is the drumstick tree, *British Medical Journal*, 331, 781

Malaria dipstick test kit 'OptiMAL-IT'. Contact: Gareth Williams, Marketing Manager, TCS Group, Botolph Claydon, Buckingham MK18 2LR, UK (Tel: +44 (0) 1296 711257 (DDI); Mob: +44 (0) 7739 236704; Fax: +44 (0) 1296 714806; E-mail: gareth.williams@tcsgroup.co.uk; www.tcsgroup.co.uk

Mavin, S., Joss, A. W. L. and Ho-yen, D. O. (2005) Incidence of Scottish Lyme disease appears to be related to the effects of weather on tick survival and human behaviour, *J Royal College of Physicians of Edinburgh*, 35, 112–15

Nathwani, D. (1998) How I manage the febrile returning traveller, *Proceedings of Royal College Physicians Edinburgh*, 28, 24–8

Philipp, R., King, C. and Hughes, A. (1992) Understanding of Weil's disease among canoeists, *British Journal of Sports Medicine*, 26, 223–7

Picozzi, K., Fevre, E.M., Odiit, M., et al. (2005) Sleeping sickness in Uganda: a thin line between two fatal diseases, *British Medical Journal*, 331, 1238–42

Pollard, A. J. and Murdoch, D. R., *The High-Altitude Medicine Handbook*, Radcliffe Medical Press Ltd, Oxford (2003) ISBN 1-85775-849-8

Tumwine, J. K. (2005) Clean drinking water for homes in Africa and other less developed countries, *British Medical Journal*, 331, 468–9

Welsby, P. D. (2005) Traveller's diarrhoea: 'Travel broadens the mind but opens the bowels', *J R Coll Phys Ed.*, 35, 40–4

GLOSSARY

Acclimation: adaptation that results from alterations to environmental conditions, for example exposure to a sauna, botanical gardens or an environmental chamber.

Acclimatization: adaptation as a result of changes to the natural environment, for example, summer weather or residence in a hot climate such as a July stay in Egypt.

Aldosterone: hormone that regulates sodium absorption.

Alveoli: sacs with thin membranes that allow gas transfer between the lungs and the blood.

Analgesics: medication to counter or reduce pain.

Antidiuretic hormone (ADH): promotes a reduced urine water loss through retention of fluid in the kidneys.

Arterial Oxygen Saturation: *see* SaO_2.

Arterioles: small arteries that carry blood to capillaries.

A-VO_2 difference: the difference between the O_2 content of the arterial and venous blood is referred to as the A-VO_2 difference (A-VO_2 diff).

Basal cell: one of the cell types in the skin. One of the types of skin cancer develops in these cells, usually as a result of too much exposure to sun.

Basal Metabolic Rate (BMR): energy expended by the body in a given time during rest in fasted conditions.

Blood: fluid that contains several parts including plasma (the watery component of blood) and red cells.

Body mass index (BMI): used to determine the degree of 'over or under weight' of a person. It is calculated by dividing body weight (kilograms) by height in meters squared. (BMI = kg/m^2).

Bone marrow: cells inside the bones that produce the red blood cells.

Bronchi: tubes that conduct air to and from lungs.

Bronchoconstrictor effect: the narrowing of the airways, as in asthma.

Bronchorrhoea: excess outpouring of secretions in the airway.

Bronchospasm: reduction in diameter of bronchi.

Capillaries: small blood vessels with very thin walls.

Cardiac arrhythmias: irregular heartbeats.

Cardiac output (CO): made up of the heart rate (the number of beats per minute) and the stroke volume (SV), the volume of blood ejected by each heart beat. At rest the CO is about 5-6ltr per minute. For example, a SV of 75ml per beat and a heart rate of 70 beats per minute results in a CO of 5.25ltr per minute.

Carotid bodies: sensors in the neck region which are stimulated by a decrease in the partial pressure of oxygen, and an increase in hydrogen ions in the arterial blood.

Chemoreceptors: sensors that detect changes of chemicals in the blood.

Cold shock response: results from sudden immersion in cold water and promotes uncontrollable hyperventilation, which may last up to 3 minutes.

Core and shell: The core comprises the brain, chest, abdominal areas and some skeletal muscles. While the core temperature is normally tightly regulated, within a narrow range, the shell temperature can vary widely, particularly the shell outer layer, the skin. The size of the shell varies depending on the environmental temperature. The shell refers to the external parts of the body. The shell's thickness varies.

Core temperature (Tc): in humans is normally around 37°C. There is diurnal variation with Tc being lowest in the morning and highest in the evening. The measurement site will influence the Tc value. The oesophageal temperature is normally higher than rectal, tympanic and oral temperatures. The oesophageal temperature responds faster to changes in core temperature than the rectal temperature. Oral temperatures can be influenced by the temperature of recently drunk fluids or ingested food. Oral and tympanic temperatures may be influenced by environmental temperatures, for example cold air or immersion in cold water.

Deep vein thrombosis: blood clots develop in the deep veins usually in the calf.

Dehydration: lowering of water volume in body.

Electrolyte: molecules such as salt (NaCl) which are dissolved in fluid water or blood.

Erythropoietin (EPO): a hormone released mainly from the kidneys. EPO is a signal for the bone marrow to produce red blood cells.

Exercise-induced bronchospasm (EIB/asthma): symptoms include breathlessness and a 'tight' chest. The incidence of EIB is three times greater in 'cold weather' athletes compared to their warm weather counterparts. Experimentally, bronchospasm can be induced in all subjects with a combination of cold air and high respiratory minute volumes. Different people have different sensitivities for the development of EIB. EIB is triggered by cooling and/or drying the lining of the airways.

Extracellular fluid: fluid that lies outside the cells, namely in the blood vessels and between the cells.

Frostbite: a localised lesion caused by freezing.

Frostnip: the skin is painful, blanches and loses sensation but remains pliable.

Gastric emptying: clearance from the stomach.

Glycogen depletion: reduced carbohydrate stores.

Glycogen resynthesis: carbohydrate ingestion after exercise that results in an increase in glycogen stores.

Haemoglobin: is contained in red cells. Haemoglobin carries almost all of the O_2 in the blood. The haemoglobin concentration for males is around 15g per 100ml of blood and for females is around 13.5g per 100ml of blood.

Heat acclimatization: results from repeated exposure to heat stress where the core temperature and sweating are increased on a regular basis.

Hydrogen ion (H^+): a positively charged atom – a proton. The H^+ concentration in a solution is a determinant of how acid the solution is.

Hyperthermia: high core temperatures.

Hypertonic: solutions with sufficient ions to have a higher osmolarity than blood.

Hyperventilation: increase in rate/depth of breathing that results in a reduction in the body's carbon dioxide stores.

Hypoglycaemia: low blood-sugar levels.

Hyponatraemia: low plasma-sodium levels.

Hypothalamus: part of the brain that influences the control of some of the internal workings of the body.

Hypothermia: low core temperature: must be below 35°C for true hypothermia to be present.

Hypotonic: Solutions that have a lower osmolarity than blood.

Interstitial fluid: fluid that surrounds the cells, that is, not in the cells and not in the blood vessels.

Intracellular fluid (ICF): in the cells.

Ion: electrically charged particle.

Jet lag: travel across time zones that upsets the normal body clock stability by producing a mismatch between the external (local) and internal (biological clock).

Kilocalorie (kcal): the amount of heat needed to increase 1ltr of water from 14.5 to 15.5°C.

Maximal oxygen uptake (VO_2 max): the greatest amount of O_2 that can be consumed per minute. Scores are given in ltr per minute or ml/kg/min when related to body weight. Typical scores for sedentary 20-year-old males and females are 45 and 40ml/kg/min respectively. Elite endurance male and female athletes have scores of ~ 80- and 70ml/kg/min respectively. See also p160.

Medulla (oblongata): respiratory control centre in the brain.

Metabolic rate: energy expended by the body in a given time.

Minute ventilation (VE): The amount of air that is moved out of the body in a minute is called the minute ventilation (VE). The amount moved in is very similar or the same as that moved out. The VE consists of the frequency of breathing (respiratory rate (RR) and the tidal volume (TV). At rest the VE is between 5 and 15ltr per minute. A resting VE of 6ltr per minute could be made up of a RR of 12 breaths per minute and a TV of 0.5ltr. See also p158.

Myocardial infarction: heart attack.

Oesophagus: tube that connects the mouth with the stomach.

Osmolarity: the concentration of solute particles in a solution. Increases in osmolarity lower the water concentration.

Osmoreceptors: detect changes in the osmolarity.

Oxygen uptake: oxygen consumed.

Partial pressure: The pressure of a given gas is directly related to the concentration of the gas. This is known as the partial pressure.

PH: a marker of the degree of acidity. A pH of 7 is neutral. As H^+ increase, the pH falls and vice-versa. A pH >7 is alkaline. A pH <7 is acid.

Plasma: the watery part of the blood, that is, all except the cells.

Raynaud's Syndrome: an abnormal arterial vasoconstriction, in response to cold stress, most commonly affecting the fingers.

Relative intensity: an example: at high altitude the same amount of work requires a higher percentage of maximal oxygen uptake than at sea level.

Respiratory rate (RR): the number of breaths per minute.

SaO_2: arterial oxygen saturation – the percentage of oxygen in the arterial blood compared with the maximum possible. The normal at sea level is 97 per cent. One hundred per cent is only reached with oxygen at high pressure as in a diving chamber.

Shivering: the involuntary contraction of a muscle group against its antagonists, for example quadriceps against hamstrings. Shivering can increase heat production 4-5 times above the normal resting level.

Sickle-cell disease: a condition with abnormal haemoglobin that, under certain conditions such as hypoxia, changes the red blood cells from smooth discs to rough sickle-shaped cells. These tend to block small blood vessels and can damage the function of many organs. The normal haemoglobin in adults is HbA. Since people get Hb genes from both parents, the normal adult haemoglobin is described as HbAA. However there are abnormal haemoglobins, HbS, the sickle-cell haemoglobin, and HbC, also found in people of negroid origin. The person with HbSS will have full blown sickle-cell problems. HbCC causes anaemia (bloodlessness) and other problems. However the combination HbSC has effects indistinguishable from the full blown sickle-cell problem. On the other hand HbSA has one normal HbA gene that partially counteracts the HbS gene and the problem is less severe.

Sleep deprivation: prolonged lack of sleep has a bad effect on an individual's performance and general well-being.

Sphincters: muscle structures which occur in a number of organs to control flow, for example there is a sphincter at the lower end of the bladder that controls the flow of urine.

Squamous cell: superficial skin cells that can develop into one form of cancer as a result of too much exposure to sun.

Stroke volume: the volume of blood ejected by each heartbeat.

Systemic: refers to the circulation of the whole body apart from the circulation in the lungs.

Thermal effectors: the thermal effector organs are the skeletal muscles, the smooth muscles of the arterioles that supply blood to the skin and sweat glands.

Thermoreceptors: skin (peripheral thermoreceptors) and central thermoreceptors: the hypothalamus, spinal cord and abdominal organs.

Tidal volume (TV): the TV is the volume of air that is moved during an inspiration or expiration.

Trachea: tubes that conduct the air to the alveoli.

Trench foot: the usual title for non freezing cold injury, develops over a fairly long period when the feet are exposed to cold (above freezing) and/or wet, for example prolonged walking in boggy ground or sweat in impervious boots.

Vasoconstriction: reduction in the diameter of blood vessels.

VE: *see* Minute ventilation.

Venous return: the blood returning to the heart through the veins.

Ventricular fibrillation: a very irregular electrical rhythm of the heart. The heart no longer pumps blood and the person is dead unless the rhythm can be corrected. One form of heart attack.

VO_2: *see* oxygen uptake.

VO_2max: *see* maximal oxygen uptake.

Zeitgebers: time-givers (German).

INDEX